DIGITAL INNOVATIONS FOR A CIRCULAR PLASTIC ECONOMY IN AFRICA

Plastic pollution is one of the biggest challenges of the twenty-first century that requires innovative and varied solutions. Focusing on sub-Saharan Africa, this book brings together interdisciplinary, multi-sectoral and multi-stakeholder perspectives exploring challenges and opportunities for utilising digital innovations to manage and accelerate the transition to a circular plastic economy (CPE).

This book is organised into three sections bringing together discussion of environmental conditions, operational dimensions and country case studies of digital transformation towards the circular plastic economy. It explores the environment for digitisation in the circular economy, bringing together perspectives from practitioners in academia, innovation, policy, civil society and government agencies. The book also highlights specific country case studies in relation to the development and implementation of different innovative ideas to drive the circular plastic economy across the three sub-Saharan African regions. Finally, the book interrogates the policy dimensions and practitioner perspectives towards a digitally enabled circular plastic economy.

Written for a wide range of readers across academia, policy and practice, including researchers, students, small and medium enterprises (SMEs), digital entrepreneurs, non-governmental organisations (NGOs) and multilateral agencies, policymakers and public officials, this book offers unique insights into complex, multilayered issues relating to the production and management of plastic waste and highlights how digital innovations can drive the transition to the circular plastic economy in Africa.

Muyiwa Oyinlola is an associate professor in Engineering for Sustainable Development and Director of the Institute of Energy and Sustainable Development at De Montfort University, UK. A chartered engineer, committed to engineering sustainable solutions for low- and middle-income countries, his work places particular emphasis on identifying and integrating socio-cultural considerations required for the long-term success of engineering projects. He leads the DITCh Plastic Network, a multi-sectoral, international and interdisciplinary network aimed at promoting and supporting digital innovations that can accelerate the circular plastic economy transition in Africa. He is committed to developing processes and products for the circular plastic economy.

Oluwaseun Kolade is a professor of Entrepreneurship and Digital Transformation at Sheffield Business School, Sheffield Hallam University, UK. He recently held academic and leadership positions at Leicester Castle Business School, De Montfort University, where he also chaired the African Entrepreneurship Cluster. With an engineering background and a PhD in International Development, Seun's research activities cover the broad areas of transformative entrepreneuring, digital transformation, circular economy, and SMEs' strategies in turbulent environments. Seun is an agile certified practitioner with experience of leading transdisciplinary projects involving international partners across public and private sectors. He has chaired international conferences and is an invited speaker in various international fora.

Routledge Studies in Sustainability

Rural Governance in the UK
Towards a Sustainable and Equitable Society
Edited by Adrienne Attorp, Sean Heron and Ruth McAreavey

The Environmental Impact of Cities
Death by Democracy and Capitalism
Fabricio Chicca, Brenda Vale and Robert Vale

Reimagining Labour for a Sustainable Future
Alison E. Vogelaar and Poulomi Dasgupta

Waste and Discards in the Asia Pacific Region
Social and Cultural Perspectives
Edited by Viktor Pál and Iris Borowy

Digital Innovations for a Circular Plastic Economy in Africa
Edited by Muyiwa Oyinlola and Oluwaseun Kolade

Critical Sustainability Sciences
Intercultural and Emancipatory Perspectives
Edited by Stephan Rist, Patrick Bottazzi and Johanna Jacobi

For more information on this series, please visit: www.routledge.com/
Routledge-Studies-in-Sustainability/book-series/RSSTY

DIGITAL INNOVATIONS FOR A CIRCULAR PLASTIC ECONOMY IN AFRICA

Edited by Muyiwa Oyinlola and Oluwaseun Kolade

Routledge
Taylor & Francis Group
LONDON AND NEW YORK

from Routledge

UK Research
and Innovation

GCRF
Global Challenges
Research Fund

Designed cover image: © kirkchai benjarusameeros / iStock

First published 2023
by Routledge
4 Park Square, Milton Park, Abingdon, Oxon OX14 4RN

and by Routledge
605 Third Avenue, New York, NY 10158

Routledge is an imprint of the Taylor & Francis Group, an informa business

British Library Cataloguing-in-Publication Data
A catalogue record for this book is available from the British Library

ISBN: 978-1-032-24410-5 (hbk)
ISBN: 978-1-032-24409-9 (pbk)
ISBN: 978-1-003-27844-3 (ebk)

DOI: 10.4324/9781003278443

Typeset in Bembo
by Newgen Publishing UK

CONTENTS

FIGURES

TABLES

CONTRIBUTORS

Soroush Abolfathi is an assistant professor and the co-chair of Global Research Priorities in Sustainable Cities at the University of Warwick, UK. His research interests include pollution transport in the environment as well as resilient infrastructures to natural hazards, climate change and flooding.

Olubunmi Ajala is a lecturer and Learning Analytics Lead at the School of Economics, Finance and Accounting, Faculty of Business and Law, Coventry University, UK, with a decade of experience in the financial sector. He is the co-author of "Understanding public sentiment in relation to African Continental Free Trade Area" and "Do lockdown and testing help in curbing COVID-19 transmission?"

Ifeoluwa Akanmu is a researcher at DITCh Plastic Network, Nigeria. She is a product manager with a Computer Science background building digital products that tackle the toughest African problems. She is a member of Women in Cleantech and Sustainability (WCS) and was selected as one of the pioneer Digital Ambassadors of African Development Bank in 2022.

Florensa Amadhila founded Epupa Cleaning Services and General Services cc Namibia in 2006, in Windhoek. She has 18 years of experience in waste management. She is passionate about waste management, quality service delivery, technology and innovation for waste management. She currently serves on the Recycling Namibia Forum Board.

Natalie Beinisch is Executive Director of Circular Economy Innovation Partnership (CEIP) in Lagos, Nigeria. Her research interests are value chain governance and transnational regulation. Prior to working on circular economy issues, she worked in both leadership development and sustainable finance.

Stuart R. Coles is Reader of Sustainable Materials and Manufacturing at WMG, University of Warwick, UK. He focuses on quantifying impacts surrounding the circular economy, the environment and sustainability, particularly when applied to the use of critical raw materials, waste products and recycled materials in industrial products and processes.

Raili Hasheela is the Co-director of Integrated Environmental Management Solutions cc in Windhoek, Namibia. She holds a PhD in Management from Azteca University, specialising in Environmental Management. She has a strong interest in integrated environmental management and sustainable development. She has coordinated several environmental management-related projects and programmes.

Celine Ilo holds an MSc in Energy Engineering and is currently a PhD researcher at the Institute of Energy and Sustainable Development (IESD), De Montfort University, Leicester, UK. She has great passion for steering positive transformations in the field of sustainable development.

Luzé Kloppers-Mouton founded The Recycling Lab cc in 2020, in Windhoek, Namibia, to contribute towards a circular economy in Namibia. She is an architect, entrepreneur and change maker, with expertise in sustainable development and project management. She currently serves on the Recycling Namibia Forum Board and works on several environmental initiatives across Southern Africa.

Oluwaseun Kolade is a professor of Entrepreneurship and Digital Transformation at Sheffield Business School, Sheffield Hallam University, UK. He recently held academic and leadership positions at Leicester Castle Business School, De Montfort University, where he also chaired the African Entrepreneurship Cluster. With an engineering background and a PhD in International Development, Seun's research activities cover the broad areas of transformative entrepreneuring, digital transformation, circular economy, and SMEs' strategies in turbulent environments. Seun is an agile certified practitioner with experience of leading transdisciplinary projects involving international partners across public and private sectors. He has chaired international conferences and is an invited speaker in various international fora.

Selma Lendelvo is an associate research professor at the Centre for Grants Management and Resource Management, at the University of Namibia, in the area of environmental and natural resource management. She is currently the

Director for Centre for Grants Management and Resource Management and has been in the research field over the past 20 years.

Sunday Augustine Leonard is a program manager at the United Nations Environment Program in the United States. He is based at the Secretariat of the Scientific and Technical Advisory Panel to the Global Environment Facility in Washington, DC, where he provides scientific and strategic advice on climate change mitigation, chemicals and waste management projects/programmes.

Victor Odumuyiwa is Director of the National Information Technology Development Agency (NITDA) IT Hub and is a senior lecturer in the Department of Computer Sciences, University of Lagos, Nigeria. He is also a member of the DITCh Plastic Network. He drives digital innovations for sustainable development and develops responsible artificial intelligence for Natural Language Processing with applications in sentiment analysis, recommender systems, cyber security and education.

Olawunmi Ogunde is a PhD student at WMG, University of Warwick, UK. Her research is focused on the plastic circular economy. She aims to explore and develop a plastic data exchange platform within the circular economy. She has years of experience on waste, recycling and promoting the sustainable development goals (SDGs).

Silifat Abimbola Okoya is a PhD researcher in the Institute of Energy and Sustainable Development at De Montfort University, UK, with a focus on UN SDG 4. She has 20 years' management experience with multinational fast moving consumer goods (FMCG) organisations in marketing, communications and development. She is the president of the Nigeria–Britain Association, an NGO fostering friendships between Nigeria and Britain.

Muyiwa Oyinlola is Director of the Institute of Energy and Sustainable Development (IESD) at De Montfort University, UK. He is an associate professor in Engineering for Sustainable Development and Chartered Engineer. He leads the DITCh Plastic Network and is involved in other initiatives aimed at tackling plastic pollution.

Mecthilde Pinto is a researcher at the University of Namibia and holds a master's degree in Tourism Studies from there. Her research focus areas and interests are in natural resource management, gender, community-based conservation, community-based tourism, waste management and several environmental aspects. She worked as the project assistant on the DITCh Plastic Project.

Barry Rawn is Associate Teaching Professor at Carnegie Mellon University Africa and Kigali Collaborative Research Centre, Rwanda. He studies the deployment of sustainable energy infrastructure in developing countries. As associate teaching professor at Carnegie Mellon University Africa, he teaches power- and energy-related courses while facilitating local industrial collaboration and international research activities. He obtained his PhD from University of Toronto, Canada, in 2010.

Chifungu Samazaka is the founder of Recyclebot, Lusaka, Zambia. He is a designer, developer and entrepreneur who is based in Lusaka. He has founded three growth-oriented high-technology start-ups. His companies include a fully distributed person-to-person waste recycling platform (Recyclebot). He is skilled in lean start-up development, technology and communications, including multiple social media outlets.

Patrick Schröder is a senior research fellow at Chatham House, Royal Institute of International Affairs, UK. His research focuses on the issues of policy, finance, trade and technology innovation in the circular economy.

John Sifani holds a PhD in Systems of Innovation from the University of Namibia. He is currently the Director of the Centre for Innovation and Development, University of Namibia. His research focuses are science and technology innovation policy, technology transfer, commercialisation of innovation and research outputs and management of innovation and research platforms.

Bosun Tijani is co-founder and CEO of Co-Creation Hub, Nigeria, and is a pan-African innovation enabler that works at the forefront of accelerating the application of innovation and social capital for a better society.

Kutoma Wakunuma is Associate Professor Research and Teaching in Information Systems at De Montfort University, UK. Her research interests are in understanding the social and ethical implications of information and communication technologies (ICTs) and the role that emerging technologies play in both the developed and developing world.

Timothy Whitehead is a senior lecturer in Product-Design at Aston University, UK, with a research interest in developing tools and approaches to improve the design of products distributed in low-income countries. He has experience in utilising design methods and technology to improve the livelihood of those living on less than $1 a day.

ACKNOWLEDGEMENTS

Firstly, we would like to acknowledge UK Research and Innovation (UKRI) for supporting this work through the Global Challenges Research Fund (GCRF) under Grant EP/T029846/1. Most of the chapters in this book resulted from the UKRI-GCRF DITCh Plastic Network.

Secondly, we would like to thank organisations, who are founding members of the DITCh Plastic Network – these include De Montfort University, Chatham House, University of Warwick, University of Birmingham, Aston University, University of Lagos, University of Namibia, Kigali Collaborative Research Centre, Co-creation Hub Lagos, Ihub, Nairobi, BongoHive Lusaka, Ocean Generation (formerly Plastic Oceans) London and Yo-Waste, Kampala.

We would also like to extend our deepest appreciation to stakeholders from across the continent who engaged with the DITCh Plastic Network through focus groups, interviews, workshops, webinars, electronic surveys and our conference. We are grateful for the support and insights you have provided, all of which we have tried to capture in this book.

Finally, we would like to thank all peer reviewers for their feedback on the chapters in this book.

1

INTRODUCTION

A Digitally Enabled Circular Plastic Economy for Africa

Muyiwa Oyinlola and Oluwaseun Kolade

1 Plastic Pollution in Africa

Plastics have been around since the discovery of polystyrene in 1839. They come in different types such as polyethylene terephthalate (PET), low-density polyethylene (LDPE), high-density polyethylene (HDPE), polyvinyl chloride (PVC) and polypropylene (PP). Between 1950 and 2017, approximately 7 billion out of 9.2 billion tonnes of plastics produced ended up as wastes. They further approximated the recycling rate to be only 9% while 79% is disposed of in landfills and the oceans. Therefore, even though plastics are essential materials due to unique properties and extensive benefits, they now pose considerable environmental and health problems due to large quantities that have been mismanaged over the years (Sakthipriya, 2022).

This challenge is expected to worsen, as several reports indicate a steady increase in annual global production of plastics. Dasgupta et al. (2022) reported a growth rate of more than 8% per year, and in 2016, Drzyzga and Prieto (2019) estimated the production to be 335 million tonnes per annum, and Plastics Europe (2021) estimated global plastic production in 2020 to be 367 million tonnes. The latest estimates at the time of writing put production at about 381 million tonnes, of which 50% is single-use (Grodzińska-Jurczak et al., 2022; Phillips, 2022). Given the current growth rate, the total plastics produced is projected to grow to 33 billion tonnes by 2050 (Rochman et al., 2013; Jambeck et al., 2015). This trend has become a major concern as plastic is non-biodegradable, and microplastics are permeating into the environment and the food chain (Wright and Kelly, 2017).

Many countries in Africa have poor infrastructure and suboptimal waste management systems, which exacerbates the plastic pollution challenge. It is estimated that less than 5% of plastic waste is recycled in Africa (UNEP, 2018)

DOI: 10.4324/9781003278443-1

while the remaining are disposed of through open dumping, open burning, unregulated landfills, and dumping into drains, streams and rivers. The scale of the challenge is expected to increase as the continent is anticipated to experience almost 200% increase in waste generated by 2050, with much of this being plastic (Kaza et al., 2018).

2 The Circular Economy as a Viable Solution

The circular economy has been touted as a viable intervention for the plastic challenge (Geissdoerfer et al., 2017; Kirchherr et al., 2017). Its key approaches of reducing, reusing, recycling, redesigning, remanufacturing and recovering are expected to significantly contribute to better management of the high volumes of waste in the ecosystem (Leslie et al., 2016). These core ideals of the circular economy have gained increasing attention globally, including interest from governments and businesses (Korhonen et al., 2018). The circular economy empowers organisations to realise lasting productivity and economic growth (Cainelli, Evangelista and Savona, 2006), otherwise described as the "circular advantage". The concept of the circular economy has been explored by several scholars such as Murray, Skene and Haynes (2017); Araujo Galvão et al. (2018); Berg et al. (2018); and Gall et al. (2020).

The circular plastic economy (CPE) is a system which employs the principles of the circular economy to the plastic value chain, including design, manufacture, use and end-of-life phase. Therefore, the CPE approach promotes innovative design, encourages recycling and incentivises the reuse of materials, thereby minimising issues arising from the use and disposal of plastic products (Völker, Kovacic and Strand, 2020). In other words, the CPE fosters a move to more sustainable interventions for the plastic challenge through innovation (Dedehayir, Mäkinen and Ortt, 2018).

While public awareness of the need for a CPE has grown, the African continent has not experienced corresponding development in terms of tangible actions and verifiable achievements. This is owing, in part, to the constraints presented by institutional frameworks in which national governments are frequently out of touch with global debate (Kolade et al., 2022), and public participation is frequently not matched by policy commitment and political resolve (Adetoyinbo et al., 2022). Furthermore, many private-sector stakeholders continue to work in silos, limiting the gains and effectiveness of current circular economy campaigns (Oyinlola et al., 2022b). According to Barrie et al. (2022), developing countries may be limited in taking advantage of the higher-value opportunities of the circular economy.

Against this backdrop, the central thesis of this book is that digital tools and technologies, which result in digital innovations (DIs), can be the game changer that positively disrupts the landscape by channelling and driving a multi-stakeholder approach that brings together digital innovators, researchers, policymakers and ordinary citizens together in the collective drive towards the CPE in Africa. DIs can facilitate the creation of new multi-sided platforms and institutions that link existing stakeholders together for greater impact. They can also enable new actors and ordinary citizens to co-create innovative solutions and ideas to drive the CPE. This book therefore explores the challenges and opportunities of a digitally enabled circular economy in Africa.

3   Digitisation in the Circular Economy

DIs create and integrate new technologies into current systems to address issues and boost productivity, accessibility, dependability and sustainability (Ciriello, Richter and Schwabe, 2018; Kohli and Melville, 2019). Internet of things (IoT), smart mobile devices, big data, remote sensing, blockchain, cloud storage, artificial intelligence (AI) and three-dimensional (3D) printing are all examples of digital tools and technology for innovation. Several business sectors in Africa have benefited from DIs. For instance, "precision agriculture" based on sensors and satellites as well as AI-based agronomic solutions have been utilised to assist sustainable agriculture in Africa, providing smallholder farmers (SHFs) and their communities with a number of advantages (Syngenta, 2019). Another industry where DIs has been effectively applied is mobile finance, which has enabled low-cost money transfers and many creative forms of financing, such as crowdsourcing and peer-to-peer lending. These tools have completely changed the African payment environment, inspiring brand-new, cutting-edge methods of approaching the financial value chain. Additionally, digital technologies such as geospatial platforms and embedded systems have transformed the energy industry in Africa by enabling real-time demand monitoring, adjustment and smarter management of distributed power systems (Annunziata et al., 2015).

There are several factors catalysing the uptake of digital tools and technologies in Africa, for example, the demographic profile of the continent; almost 60% of Africa's population is under the age of 25 (Statista, 2021). Furthermore, Africa has the fastest growing internet penetration (GSMA, 2020; Granguillhome Ochoa et al., 2022), and the continent has attracted significant investment in digital platforms such as Google AI hub in Ghana and Facebook hub in Kenya. In addition, several technology innovation hubs have sprung up across the continent (Atiase, Kolade and Liedong, 2020), giving several young people the opportunity to immerse themselves in technologies which result in innovations that can support development. The mushrooming of these tech-hubs, which offer space and technology support for budding digital entrepreneurs, is empowering young Africans to be more creative and more innovative in their use of DI. According to GSMA (Giuliani and Ajadi, 2019), the tech-hubs offer support as incubators, accelerators, university-based innovation hubs, maker spaces, technology parks and co-working spaces. The tech-hubs are instrumental in building DI start-ups and a robust digital ecosystem where entrepreneurs can learn from as well as share ideas with like-minded innovators. Furthermore, tech-hubs offer much-needed fast internet access and electricity (Giuliani and Ajadi, 2019).

DIs have also demonstrated the ability to contribute to Africa's CPE by filling the vacuum left by insufficient waste collection and management infrastructure (Antikainen, Uusitalo and Kivikytö-Reponen, 2018). Several improvements have been implemented in a bid to transform the plastic value chain into a smart, innovative and sustainable value network through improved plastic identification, collection, transportation, sorting, processing and reuse. Alternative Energy Solutions (AES) in Kenya, for example, uses revolutionary technology to turn various types of plastic into oil (Horvath, Mallinguh and Fogarassy, 2018). Pratap et al. (2019) propose an automated communication method based on IoT technology between households and waste collection agencies to help monitor and collect plastic waste, recycle and aid in centralised disposal. As IoT becomes more ubiquitous on the African continent, such a model could also be considered among other DIs. Mugo and Puplampu (2020) discuss how smart sensors as a technology innovation can be useful in addressing environmental pollution and waste management in Africa. Singh (2019) discussed how municipal waste management can make use of geographic information systems (GIS) and the layers available from remote sensing. Chidepatil et al. (2020) posit that AI and blockchain technology have the potential to make recycling more efficient. They suggest that using AI to segregate plastic waste will ensure effective and intelligent segregation, which could otherwise be a complex and inefficient procedure. Furthermore, they suggest that blockchain technology can be utilised as a "trust-based platform between plastic waste segregators, recyclers, and recycled feedstock buyers (manufacturers)", so that information can be easily exchanged and validated between the various partners in the value chain, making it easy for partners to have relevant information on plastic waste and how best to reduce or recycle it. There are several start-ups utilising digital technology to tackle the plastic pollution challenge in Africa, Figure 1.1 presents some of these, and a comprehensive list is presented in Oyinlola et al. (2022).

FIGURE 1.1 Some start-ups utilising digital technologies for the circular plastic economy

4 DITCh Plastic Project

Despite a growing number of studies and start-ups focused on utilising digital tools and technologies for the CPE over the past decade. Progress has been slow, with most start-ups struggling to scale. In response to this, the United Kingdom Research and Innovation (UKRI), through the Global Challenges Research Fund (GCRF), funded the formation of a network in 2020, https://gtr.ukri.org/proje cts?ref=EP%2FT029846%2F1. The Digital Innovations for Transitioning to a Circular Plastic Economy in Africa (DITCh Plastic) Network, led by the partners in Figure 1.2, is a multisectoral, international and interdisciplinary network aimed at promoting and supporting DIs that can accelerate the transition to a CPE in Africa. The network targeted to characterise, cluster, synergise and optimise DIs that would support a transition to a CPE in Africa and had the following specific objectives:

I. Identify and assess digital solutions and innovations that can support the transition to a CPE.
II. Characterise technical, political, gender, socioeconomic and cultural factors that can influence the transition to a CPE.
III. Identify policy, research questions and capacity building opportunities for promoting digital tools and innovations.
IV. Promote digital tools and innovations for a CPE.

This book, in addition to Kolade et al. (2022), Oyinlola et al. (2022, 2022b) and Schroeder et al. (2023), is an output from the network. This book brings together interdisciplinary, multisectoral and multi-stakeholder perspectives exploring challenges and opportunities of utilising DIs to manage and accelerate the transition to a CPE in Africa. It provides both scholarly and practitioner perspectives on the role of DIs, such as web-based/mobile apps, blockchains and 3D printing, in the drive towards the CPE in Africa. These are reinforced with

FIGURE 1.2   DITCh plastic network partners (http://ditch-plastic.org/)

real-world examples, policy insights and country case studies spanning Western, Eastern and Southern Africa regions.

Along with a critical synthesis of the extant literature, DITCh Plastic has also engaged with hundreds of stakeholders from across the continent through focus groups, interviews, workshops and a conference as well as a cross-sectional survey of 1475 households across five countries which were selected to have a comprehensive representation of sub-Saharan Africa. Geographic diversity was ensured by a wide continental coverage (East, West and South): significant differences in economy size [Nigeria with a gross domestic product (GDP) of $375.8 billion versus Rwanda with a GDP of $9.137 billion], population (190 million in Nigeria to 2.5 million in Namibia) and literacy rates. Below are some insights from the extensive engagement activities.

Firstly, despite the numerous prospects in the sector, funding appears to be a significant barrier. This includes a lack of financing and/or a lack of understanding of funding sources to support research, innovation and development for the CPE. Although waste management initiatives are generally recognised as viable businesses in the medium to long term, start-ups in this sector are finding it difficult to access start-up capital to pilot their innovation until they can make a viable investment case for scaling. Digital platforms can be used to create virtual marketplaces which streamlines and optimises the plastic value chain. Government's ambitions should be to maximise the recycling (and recovery) of waste resources for productive use while reducing pollution. Despite creating online marketplaces being simple, a critical research question is how it functions with or without market regulation or support, and how this can promote good waste commodity governance.

Secondly, regulation is a significant obstacle that may slow the rate of transition. Addressing serious deficiencies and flaws in rules and regulations for sustainable plastic waste management is critical. This encompasses policy design, implementation and enforcement. It was observed that there are many excellent waste management policies across the continent; however, the majority of them might benefit from better coordination and enforcement. One example is the extended producer responsibility (EPR) scheme, in which plastic manufacturers contribute to post-consumer recovery. Another piece of helpful regulation will be on recycled content; for example, requiring plastic manufacturers to include recycled content will raise demand while also increasing recycling rates. Government incentives will be critical to success; therefore, striking the correct balance between rewards and penalties is critical. Governments must regularly examine their policies and procedures for effective waste management in their communities and industries.

It should also be noted that due to the intricacies of waste management as a regulated business with significant material flows, data will always be a critical requirement if the system is to perform properly. The lack of good data is now a major impediment to the transition to a CPE. It is difficult for industry participants and stakeholders to control something they cannot quantify. As a result, systems and/or technologies for collecting and tracking waste data must be established. Types, location, distribution, quantity collected, quantity recycled and so on should all be included in this data. Digital technologies and advances can significantly aid in the collection and analysis of pertinent data, allowing for more successful research.

There is an urgent need for public awareness/education on sustainable waste management, particularly among youth, who constitute a substantial proportion of the population. Currently, most of the population does not consider waste to be a resource, and some collectors are still struggling to obtain enough plastics for recycling. Citizen education, community awareness activities, and behavioural change programmes will need to be established and implemented. These should be implemented in cooperation with relevant policies.

The transition to a CPE demands cross-sectoral cooperation. Collaboration and coordination among various stakeholders is currently minimal, which is a major hurdle in achieving considerable progress. There must be multi-stakeholder collaboration and synergy involving the government, corporations, universities, civil organisations, local governments and communities in both urban and rural areas. National platforms that can support this type of involvement are desperately needed. This could solve the issue of coordinating many stakeholders – waste producers, waste collectors, consumers and ministries – who may need to collaborate, sometimes outside of their apparent areas of responsibility.

Digital tools and technology can be utilised to scale up local projects at granularities appropriate for Africa's population and terrain, from rural to urban communities. DIs such as mobile applications can aid in the effective collecting and transport of waste plastics to aggregators, as well as technologies that allow for the efficient optical sorting of plastics to meet reprocessing requirements. This facilitates the "bottom-up" approach to waste management and the advancement of local pollution control strategies. However, the primary potential of DIs will be in the ability to support a systemic shift to circularity at scale. Taking this forward will require further study, which will be enabled by the creation of spatial and temporal data, which informs the assessment of the systems and processes underlying waste plastic management.

A considerable demand exists to develop skills that are relevant to the circular economy. For example, introducing training in various waste management methods, including behaviour interventions, can have a significant influence within and beyond industry sub-sectors.

Another issue that must be addressed is the sociocultural dynamics of waste management. It has been discovered, for example, that social standing influences how people approach reuse and recycling. Another important difficulty is stigma; plastic waste collection is primarily viewed as a dirty job for the poor. To successfully transition, stereotypes and stigma about waste management need to be eliminated.

While both genders actively participate in the CPE, it seems like macro-level projects/initiatives are mainly dominated by males and micro-level projects by females. Similarly, while women are highly involved on the ground, they have limited opportunities in the decision-making processes for policies and strategies.

These observed gender differences need to be addressed, and there should be mainstreaming of gender-balanced projects.

Across the continent, it seems like the invitation to invest in alternative packaging has not really gained much traction in spite of several innovations around the continent. For example, biodegradable packaging has been produced locally using banana leaves and water hyacinth. However, this and similar innovations have not diffused across the continent. Systems need to be in place to promote, celebrate and diffuse these sorts of innovations across the continent.

Finally, we observed that the majority of the initiatives and interventions are focused on recycling, i.e. collection and sorting, with not very much in terms of preventing plastic waste. Therefore, there needs to be increased activity on reducing and reusing plastics.

5   Introduction to Book Chapters

The issues in this book are explored within the framework of three thematic sections: the environment for digitisation in the circular economy; digitisation in action; a digitally enabled CPE. In Chapter 2, from a multilateral agency perspective, Leonard kicks off the first section on the environment for digitisation by discussing the barriers and enabling conditions across the regulatory and institutional; economic and financial; technology and capacity; and societal and cultural dimensions. The chapter illuminates the environment that needs to be in place for a successful CPE transition. It further highlights the importance of the systems thinking approach in developing solutions and the need for the government to play a leading role in this transition. In Chapter 3, Beinisch examines the sustainable plastics regime complex – an array of partially overlapping and non-hierarchical institutions governing a particular issue area (Raustiala and Victor, 2004). She highlights that transition to a CPE is a regime complex which involves national regulators, multilateral institutions, civil society organisations and advocacy networks, market-based regulators, multinational businesses, entrepreneurs and academia. She examines how Nigerian organisations are participating in the regime complex for sustainable plastics and highlight opportunities to use it to build local institutional strength. A successful transition to a CPE requires an understanding of plastic value chains. In Chapter 4, Schröder and Oyinlola provide an overview of the plastic value chain in Africa and illustrate how digital tools and technologies can help in minimising leakage and improving material flow in the value chain. They argue that a life cycle perspective and understanding of the plastic value chains from production to end of life is fundamental to finding systemic solutions for a CPE.

The first section is concluded with Chapter 5, where Tijani, Oyinlola and Okoya utilise a practitioner's perspective along with the sectoral systems of innovation framework to examine the CPE innovation ecosystem in Africa. They postulate that the CPE ecosystem is driven by a set of local and international actors, networks and institutions, which include development organisations like the Africa Development Bank (AfDB), civil society organisations, research institutes, academic institutions, innovation intermediaries like technology hubs, investors and entrepreneurs. They propose a process that involves systematic interactions among a wide variety of actors to drive progress, activities, and the generation of knowledge relevant to innovation.

In recognition of the heterogenous, culturally and politically diverse nature of African states, Section 2, Digitisation in Action, explores specific country, regional and digital technology case studies in relation to the development and implementation of different innovative ideas to drive the CPE on the continent. The case studies discussed represent diverse socioeconomic, cultural, geographical and political landscapes, in order to adequately illuminate contextual peculiarities and common theoretical and practical insights that can inform policy and practice. The section opens with Chapter 6, where Oyinlola, Okoya and Whitehead focus on additive manufacturing, also known as 3D printing, which has been recognised as a leading frontier technology that has a significant role to play in international development (Ramalingam et al., 2016). They illustrate through case study examples how local plastic waste can be converted into filament for 3D printing and used in the creation of new, innovative, locally made products which meet specific local needs. They further highlight that utilising this frontier technology (3D printing) can result in leapfrogging traditional manufacturing, which is highly capital intensive, and the technology has the ability to create new businesses and support wealth generation. In Chapter 7, Kolade continues the discussion on plastic value chains, with a focus on blockchains. He reviews the relevance and application of blockchains in the circular economy. Utilising BanQu (a blockchain solution launched in partnership with Coca-Cola Africa to improve local recycling and drive a CPE in South Africa) as a case study, he discusses the distinct set of possibilities provided by blockchains to drive a major shift in thinking and approach. He opines that blockchains can drive a major shift in perception of plastics from wastes to assets and incentivise different behaviours by offering users the opportunities to capture value from end-of-life plastic products. He further argues that adopting blockchain in the plastic value chains in Africa can offer a more transparent and accountable system whereby information from the "molecular barcode" of plastics can be publicly tagged and tracked, but not altered, through the product life cycle. In Chapter 8, Odumuyiwa and Akanmu discuss initiatives and interventions using digital tools/innovations to tackle the plastic waste challenge in West Africa. They highlight various examples of how DIs have been used to advance the CPE in West Africa. They further identify the gaps that need to be addressed. In Chapter 9, Kolade, Oyinlola and Rawn draw on in-depth interviews and focus group discussions with key stakeholders to examine the many threads held by researchers, entrepreneurs and industrialists, investors and policymakers in East Africa. They explicate the collaborative synergy of stakeholders across sectors that play a critical role in the transition to a CPE in

East Africa. They also highlight the important contributions of DIs in lowering barriers and changing attitudes among consumers and producers alike. Section 2 ends with Chapter 10, where Lendelvo, Pinto, Amadhila, Kloppers, Samazaka, Hasheela and Sifani discuss case studies from Southern Africa. They draw on cross-sectional engagement with stakeholders to highlight six opportunity areas/drivers for DI and the use of technology for the CPE, including environmental sustainability, technological and DIs, economic significance, employment creation and enterprise, livelihood improvement and gender equality.

The final section of this book draws on contributions from both academia and practitioners to make proposals for a digitally enabled CPE. In Chapter 11, Ilo, Oyinlola and Kolade draw on the extant literature to propose the BIG-STREAM framework, which highlights digital functions and strategies for a digitally enabled CPE. They highlight big data, AI, IoT, mobile applications, GIS and remote sensing as critical digital functions. In Chapter 12, Ogunde, Oyinlola, Coles make a contribution to the discourse on the global plastic crisis with particular emphasis on how plastic management in Africa can be enhanced with adequate data. They highlight that effective data collection and usage will be facilitated by a multi-stakeholder, multi-process and multisectoral approach and, therefore, argue for a plastic data exchange (PDE) platform which will facilitate collaboration between stakeholders. In Chapter 13, Okoya, Oyinlola, Schröder, Kolade and Abolfathi investigate how small- and medium-sized enterprises (SMEs) are utilising digital technology for decentralised plastic waste management solutions. They showcase case studies from around the continent and emphasise which technologies are currently employed. They observe that these start-ups' activities are focused on one or more of three key areas: subscription, collection, and processing. They add that the decentralised method used across Africa provides considerable social, environmental and economic benefits to stakeholders.

In Chapter 14, Ajala utilises machine learning for text analysis of policy description. He finds the continent's efforts ineffective at directing the continent towards a circular economy due to shallow regulations, exclusion of the informal recycling sector, enforcement problems, and lack of awareness of policies, among others. He presents some broad propositions on how digital and technological tools can be used to redirect the continent from linear to circular economy and how they can also aid in plastic waste policy effectiveness.

In Chapter 15, Wakunuma and Lendelvo interrogate the gender inequalities in the CPE and examine how DI can reduce these disparities to provide opportunities for both men and women to participate and benefit equally. They note that although some innovative approaches to the CPE have been initiated by women, generally more women still work in the lower echelons of the CPE as plastic waste pickers. They further discuss how the gender gap could be reduced when looking at DI in the CPE in Africa. They propose a gender mainstreaming approach which will result in an informative and transformative change in the CPE in as far as gender and DI are concerned.

Overall, the chapters across these three sections offer a unique insight into complex, multilayered issues of transitioning to the CPE and highlight how DIs can drive the transition to the CPE in a continent where progress has been decidedly slow. As well as identifying threads of common challenges and practices, this book weaves a promising narrative of a circular economy powered by an integrated combination of DIs, policy innovations and market processes.

Acknowledgement

This work was partly supported by the UKRI GCRF under Grant EP/T029846/1.

References

Adetoyinbo, A. *et al.* (2022) 'The role of institutions in sustaining competitive bioeconomy growth in Africa – Insights from the Nigerian maize biomass value-web', *Sustainable Production and Consumption*, 30, pp. 186–203. doi: 10.1016/J.SPC.2021.11.013

Annunziata, M. *et al.* (2015) *Powering the Future: Leading the Digital Transformation of the Power Industry*. Available at: www.res4med.org/wp-content/uploads/2017/05/digital-energy-transformation-whitepaper.pdf (Accessed 13 May 2021).

Antikainen, M., Uusitalo, T. and Kivikytö-Reponen, P. (2018) 'Digitalisation as an enabler of circular economy', *Procedia CIRP*, 73, pp. 45–49. https://doi.org/10.1016/j.procir.2018.04.027

Araujo Galvão, G. D. *et al.* (2018) 'Circular economy: Overview of barriers', *Procedia CIRP*, 73, pp. 79–85. doi: 10.1016/j.procir.2018.04.011

Atiase, V. Y., Kolade, O. and Liedong, T. A. (2020) 'The emergence and strategy of tech hubs in Africa: Implications for knowledge production and value creation', *Technological Forecasting and Social Change*, 161, p. 120307. doi: 10.1016/J.TECHFORE.2020.120307

Barrie, J. *et al.* (2022) 'The circularity divide: What is it? And how do we avoid it?' *Resources, Conservation and Recycling*, 180, p. 106208. doi: 10.1016/J.RESCONREC.2022.106208

Berg, A. *et al.* (2018) *Circular Economy for Sustainable Development*. Finnish Environment Institute.

Cainelli, G., Evangelista, R. and Savona, M. (2006) 'Innovation and economic performance in services: A firm-level analysis', *Cambridge Journal of Economics*, 30(3), pp. 435–458.

Chidepatil, A. *et al.* (2020) 'From trash to cash: How blockchain and multi-sensor-driven artificial intelligence can transform circular economy of plastic waste?' *Administrative Sciences*, 10(2), p. 23.

Ciriello, R. F., Richter, A. and Schwabe, G. (2018) 'Digital innovation', *Business & Information Systems Engineering*, 60(6), pp. 563–569.

Dasgupta, S., Sarraf, M. and Wheeler, D. (2022) 'Plastic waste cleanup priorities to reduce marine pollution: A spatiotemporal analysis for Accra and Lagos with satellite data', *Science of the Total Environment*, 839(March), p. 156319. doi: 10.1016/j.scitotenv.2022.156319

Dedehayir, O., Mäkinen, S. J. and Ortt, J. R. (2018) 'Roles during innovation ecosystem genesis: A literature review', *Technological Forecasting and Social Change*, 136, pp. 18–29.

Drzyzga, O. and Prieto, A. (2019) 'Plastic waste management, a matter for the "community"', *Microbial Biotechnology*, 12(1), p. 66.

Gall, M. *et al.* (2020) 'Building a circular plastics economy with informal waste pickers: Recyclate quality, business model, and societal impacts', *Resources, Conservation and Recycling*, 156(September 2019), p. 104685. doi: 10.1016/j.resconrec.2020.104685

Geissdoerfer, M. *et al.* (2017) 'The circular economy – A new sustainability paradigm?' *Journal of Cleaner Production*, 143, pp. 757–768. doi: 10.1016/J.JCLEPRO.2016.12.048

Giuliani, D. and Ajadi, S. (2019) *618 Active Tech Hubs: The Backbone of AFRICA'S Tech Ecosystem, Mobile for Development*. Available at: www.gsma.com/mobilefordevelopm ent/blog/618-active-tech-hubs-the-backbone-of-africas-tech-ecosystem/ (Accessed 14 May 2021)

Granguillhome Ochoa, R. *et al.* (2022) 'Mobile internet adoption in West Africa', *Technology in Society*, 68, p. 101845. doi: 10.1016/J.TECHSOC.2021.101845

Grodzińska-Jurczak, M. *et al.* (2022) 'Contradictory or complementary? Stakeholders' perceptions of a circular economy for single-use plastics', *Waste Management*, 142, pp. 1–8. doi: 10.1016/J.WASMAN.2022.01.036

GSMA (2020) *The Mobile Economy*. Available at: www.gsma.com/mobileeconomy/# (Accessed 14 May 2021).

Horvath, B., Mallinguh, E. and Fogarassy, C. (2018) 'Designing business solutions for plastic waste management to enhance circular transitions in Kenya', *Sustainability*, 10(5), p. 1664. https://doi.org/10.3390/su10051664

Jambeck, J. R. *et al.* (2015) 'Plastic waste inputs from land into the ocean', *Science*, 347(6223), pp. 768–771.

Kaza, S. *et al.* (2018) *What a Waste 2.0: A Global Snapshot of Solid Waste Management to 2050*. World Bank Publications.

Kirchherr, J., Reike, D. and Hekkert, M. (2017) 'Conceptualizing the circular economy: An analysis of 114 definitions', *Resources, Conservation and Recycling*, 127(April), pp. 221–232. doi: 10.1016/j.resconrec.2017.09.005

Kohli, R. and Melville, N. P. (2019) 'Digital innovation: A review and synthesis', *Information Systems Journal*, 29(1), pp. 200–223.

Kolade, O. *et al.* (2022) 'Technology acceptance and readiness of stakeholders for transitioning to a circular plastic economy in Africa', *Technological Forecasting and Social Change*, 183, p. 121954. doi: 10.1016/J.TECHFORE.2022.121954

Korhonen, J. *et al.* (2018) 'Circular economy as an essentially contested concept', *Journal of Cleaner Production*, 175, pp. 544–552. doi: 10.1016/J.JCLEPRO.2017.12.111

Leslie, H. A. *et al.* (2016) 'Propelling plastics into the circular economy – Weeding out the toxics first', *Environment International*, 94, pp. 230–234. doi: 10.1016/ J.ENVINT.2016.05.012

Mugo, S. M. and Puplampu, K. P. (2020) 'Scientific innovations and the environment: Integrated smart sensors, pollution and E-waste in Africa'. In: Arthur, P., Hanson, K., Puplampu, K. (eds) *Disruptive Technologies, Innovation and Development in Africa*. International Political Economy Series. Palgrave Macmillan, Cham. https://doi. org/10.1007/978-3-030-40647-9_4.

Murray, A., Skene, K. and Haynes, K. (2017) 'The circular economy: An interdisciplinary exploration of the concept and application in a global context', *Journal of Business Ethics*, 140(3), pp. 369–380. doi: 10.1007/s10551-015-2693-2

Oyinlola, M. *et al.* (2022) 'Digital innovations for transitioning to circular plastic value chains in Africa', *Africa Journal of Management*, 8(1), pp. 83–108. doi: 10.1080/ 23322373.2021.1999750

Oyinlola, M., Kolade, O., Schroder, P., Odumuyiwa, V., Rawn, B., Wakunuma, K., Sharifi, S., Lendelvo, S., Akanmu, I., Mtonga, R., Tijani, B., Whitehead, T., Brighty, G., Abolfathi, S. (2022b). A socio-technical perspective on transitioning to a circular plastic economy in Africa. *SSRN Electronic Journal.* https:// doi.org/ 10.2139/ ssrn.4332 904.

Phillips, E. (2022) *100+ Plastic in the Ocean Statistics and Facts 2021–2022.* Available at: www. condorferries.co.uk/plastic-in-the-ocean-statistics (Accessed 25 September 2022).

Plastics Europe (2021) *Plastics the Fact 2021, Plastics Europe Market Research Group (PEMRG) and Conversio Market & Strategy GmbH,* 1–34. https://plasticseurope.org/ (Accessed 25 September 2022).

Pratap, A. *et al.* (2019) 'IoT based design for a smart plastic waste collection system', *Proceedings of International Conference on Sustainable Computing in Science, Technology and Management (SUSCOM),* Amity University Rajasthan, Jaipur - India, February 26-28, 2019, Available at SSRN: http://dx.doi.org/10.2139/ssrn.3356267

Ramalingam, B. *et al.* (2016) *Ten Frontier Technologies for International Development.* IDS. Available at: https://assets.publishing.service.gov.uk/media/58483675ed915d0b12000 05b/FrontierWEB.pdf (Accessed 28 September 2019).

Raustiala, K. and Victor, D. G. (2004) 'The regime complex for plant genetic resources', *International Organization,* 58(2), pp. 277–309.

Rochman, C. M. *et al.* (2013) 'Classify plastic waste as hazardous', *Nature,* 494(7436), pp. 169–171.

Sakthipriya, N. (2022) 'Plastic waste management: A road map to achieve circular economy and recent innovations in pyrolysis', *Science of the Total Environment,* 809, p. 151160. doi: 10.1016/j.scitotenv.2021.151160

Schroeder, P. *et al.* (2023) 'Making policy work for Africa's circular plastics economy', *Resources, Conservation & Recycling,* 190, p. 106868. doi: 10.1016/j.resconrec.2023.106868

Singh, A. (2019) 'Remote sensing and GIS applications for municipal waste management', *Journal of Environmental Management,* 243, pp. 22–29. doi: 10.1016/j.jenvman.2019.05.017

Statista (2021) *Africa: Population by Age Group 2020, Population of Africa 2020, by Age Group.* Available at: www.statista.com/statistics/1226211/population-of-africa-by-age-group/ (Accessed 26 September 2021).

Syngenta (2019) *How Can Digital Solutions Help to Feed a Growing World?* Available at: www. syngentafoundation.org/file/12811/download (Accessed 13 May 2021)

UNEP (2018) *Africa Waste Management Outlook..* Available at: www.unep.org/ietc/resour ces/publication/africa-waste-management-outlook (Accessed 13 May 2021)

Völker, T., Kovacic, Z. and Strand, R. (2020) 'Indicator development as a site of collective imagination? The case of European Commission policies on the circular economy', *Culture and Organization,* 26(2), pp. 103–120. doi: 10.1080/14759551.2019.1699092

Wright, S. L. and Kelly, F. J. (2017) 'Plastic and human health: A micro issue?' *Environmental Science & Technology,* 51(12), pp. 6634–6647.

PART I

The Environment for Digitisation in the Circular Plastic Economy

2
ENABLING A SUCCESSFUL TRANSITION TO A CIRCULAR PLASTIC ECONOMY IN AFRICA

Sunday Augustine Leonard

1 The plastic pollution challenge

Plastic products have been very beneficial to human development and can be considered one of the world's greatest innovations and most-used materials. However, the continued unsustainable consumption of plastics has become a source of adverse environmental impacts and negative human health effects. This is because of the sheer scale of their production and use and a lack of good post-use management practices globally (WEF, 2016). Plastic production processes, including the use of non-regenerative virgin fossil-fuel feedstock, consumption patterns and poor end-of-life management practices, have made plastics a significant contributor to greenhouse gas emissions (and consequently global warming), fresh and marine water pollution, biodiversity loss, land degradation and chemical contamination (Barra and Leonard, 2018).

The core driver of the adverse impacts summarised above is the current plastic production and consumption approach, which is mainly a linear take, make, use and dispose model. Contrary to the facts, this model assumes that resources are infinite, and the earth system is resilient and has an unlimited assimilative capacity to withstand the harmful effects of human activities. But scientific findings (for example, Steffen et al., 2015; Persson et al., 2022) have shown the limits of the earth systems. Hence, to solve the plastic pollution challenge, adopting a new model that promotes the efficient use of resources and considers the earth system's finite resilience and assimilative capacity through a circular economy approach is essential.

DOI: 10.4324/9781003278443-3

2 The circular plastic economy

The circular economy provides a holistic and systematic approach to addressing the adverse effects of plastic production and consumption. It is

> an alternative to the current linear, take, make, use, dispose economy model, that aims to keep resources in use for as long as possible, extract the maximum value from them while in use, and recover and regenerate products and materials at the end of their service life, thereby promoting a production and consumption model that is restorative and regenerative by design.
>
> *Barra and Leonard, 2018*

According to the WEF (2016), delivering the circular economy in the plastic sector would require improving the economic viability of recycling and reuse of plastics; halting the leakage of plastics into the environment, especially waterways and oceans; and decoupling plastic production from fossil-fuel feedstocks, while embracing renewable feedstocks. Implementing these strategies will require the producers, consumers, government and other stakeholders involved in plastic manufacture, use and management to adopt circular principles and work together, as depicted in Figure 2.1 and described below.

CIRCULAR ECONOMY SOLUTIONS	DESCRIPTION	AFRICAN EXAMPLES
PLASTICS FROM ALTERNATIVE FEEDSTOCKS	Produce plastics from alternative feedstocks including bio-based sources such as sugarcane, oils and cellulose, as well as from greenhouse gas, sewage sludge, food waste and natural occurring biopolymers	• TexFad (Uganda) produces carpets from banana plant materials (https://texfad.co.ug/banana-project.php) • Hya Bioplastics (Uganda) produces packaging products from cassava starch and banana stem (https://hyabioplastics.com/) • CSIR (South Africa) produces 100% biodegradable and compostable plastics (https://www.csir.co.za/bioplastic-technology)
REDESIGN PLASTIC PRODUCTS	Design plastic products to enhance longevity, reusability, recycling and waste prevention, by deploying green and sustainable chemistry and incorporating after-use, asset recovery, harmful chemical avoidance, and waste and pollution prevention from the onset	• TexFad, Hya Bioplastic, CSIR are examples of redesigning plastics • The South Africa Plastic Pact aims to redesign problematic and unnecessary plastic packaging (https://www.saplasticspact.org.za/how/) • 3D printed plastics – Kenya, Rwanda and Nigeria (Magoum, 2021)
SUSTAINABLE BUSINESS MODELS	Implement business models that promote products as services and encourage product sharing, leasing and takeback, thereby optimizing product utilization and decreasing volume of manufactured goods	• Maji Jibu Company Ltd (Tanzania) provides access to safe and affordable drinking water through a decentralized refill business model that allow 20-liter water containers to be reused continually (Footprints Africa, 2021)
BUSINESSES AND CONSUMER COOPERATION	Business to business cooperation and collaboration between businesses and consumers, whereby by-products or waste from one industry or consumers become raw material for producing new products. This may include industrial symbiosis, urban-industrial symbiosis and urban mining	• The Western Cape Industrial Symbiosis Program in South Africa facilitates cooperation between member companies resulting in exchange and reuse of plastics (Alli and Leonard, 2021) • Mr. Green Africa connects with informal waste pickers to collect and recycle plastics into valuable materials (Footprint Africa, 2021)
PLASTIC WASTES AS RESOURCE	Using end of life plastics for the remanufacturing of new plastics, for example, through chemical recycling or upcycling – i.e., conversion into other valuable products	• Plastics wastes for road construction, Ghana (Appiah et al., 2017) • Mr. Green Africa connects with informal waste pickers to collect and recycle plastics into valuable materials (Footprint Africa, 2021) • Pyramid Recycling Ghana covert plastic waste to curtain ropes, chair fittings, and wood plastics (Footprints Africa, 2021) • Recycled plastic brick factory in Cote d'Ivoire (UNICEF, 2019)
ROBUST INFORMATION PLATFORMS	Robust information platforms linking industries as well as consumers to ensure the flow of data and information on plastics.	• The Ghanian Waste Recovery Platform (https://ghanawasteplatform.org/) • WeCyclers recycling platform, Nigeria (https://www.wecyclers.com/) • Yo-Waste platform, Uganda (https://yowasteapp.com/)

FIGURE 2.1 Circular plastic economy solutions to address the plastic pollution challenge. Adapted from Barra and Leonard (2018), with the addition of specific examples from Africa

In a circular plastic economy, the use of non-toxic and renewable alternative feedstocks and energy in production must be prioritised. This may involve using bio-based feedstock sources such as oils, starch, sewage sludge and food wastes for plastic production instead of non-regenerative virgin fossil-fuel feedstock (for example, Reddy et al., 2013; Hatti-Kaul et al., 2020). Waste CO_2 and methane have also been trialled and could serve as alternative feedstock that can concurrently mitigate greenhouse gas emissions (Peach, 2014; Gu et al., 2021; IEF, 2021; Krymowski, 2021). Further, the design of plastics and associated products in a circular economy must adopt a complete life cycle perspective that promotes the use of the right and non-toxic materials and products designed for an appropriate lifetime and extended future use, including ease of reuse and recycling. This includes using eco-friendly additives that could eliminate harmful chemicals in plastic manufacture, for example, through green chemistry (Beach et al., 2013; Papaspyrides and Kiliaris, 2014).

At the end of life of plastic products, the circular economy ensures appropriate end-of-life options that allow the waste to be used as resources with priority for upcycling the materials. Examples of the recovery and conversion of waste plastics into new value products include making bricks and composite (e.g., Ahmetli et al., 2013; Guzman and Munno, 2015; Lundquist et al., 2020) in road construction (Khan et al., 2016; Appiah et al., 2017), making fabrics and other textiles (e.g., Tshifularo and Patnaik, 2020; Alberghini et al., 2021; Sadeghi et al., 2021) and producing new plastics or other chemicals through chemical recycling by breaking down into chemical component (e.g., Panda et al., 2010; Khoonkari et al., 2015; Rahimi and Garcia, 2017; Thiounn and Smith, 2020).

Essential for a successful transition to a circular plastic economy is increased cooperation between businesses and with consumers to facilitate the continued use of plastics in the economy at their maximum value through processes such as industrial symbiosis and urban mining (e.g., Sun et al., 2016; Marinelli et al., 2021). Coupled with this is embracing new business models that build on the interaction between products and services and promote efficient resource use to create greater product value, such as the products-as-a-service, sharing economy, reverse logistics and product takeback models (WEF, 2016). And this will need to be supported by incorporating digital innovation and robust information exchange platforms to help track and optimise resource use and strengthen communication and collaboration across the plastic value chain, including with consumers and other stakeholders (e.g., Oyinlola et al., 2022).

The need to adopt the circular economy approach in the plastic sector is becoming mainstream, and some initiatives towards this are occurring in Africa. For example, plastic is being produced from biological feedstocks, e.g., maize husk, sugar cane and water hyacinth in Uganda (Footprints Africa, 2021). The MARPLASTICs and other Eastern, Southern and Western African projects have helped develop plastic waste collection and upcycling solutions (Veolia, 2019; Footprints Africa, 2021; IUCN, 2021a). New circular business models that ensure that durable plastics remain in the economy for as long as possible through a decentralised franchise model for drinking water provision have been implemented in Tanzania (Footprints Africa, 2021). The use of waste plastic materials for road construction has also been demonstrated in Ghana (Appiah et al., 2017). Also, in Ghana, the Waste Recovery Platform (https://ghanawasteplatform.org/) connects stakeholders, including industries, civil societies and government actors, to facilitate waste management and support policy implementation (UNDP Ghana, 2020). These examples and others are highlighted in column 4 of Figure 2.1.

However, the technical aspects of the circular plastic solutions alone are not enough to achieve the desired transition. The plastic pollution problem persists, and there is a minimal change in the effective use of resources in the plastic sector (Jambeck et al., 2018; Babayemi et al., 2019; Ayeleru et al., 2020). Harmful chemicals are still being used in plastics, and the sector continues to be a substantial contributor to greenhouse gas emissions, ecosystem damage and adverse human health effects (WEF, 2016; PEW, 2020; Kelleher, 2021). Current solutions will not be enough to change the data presented in Jambeck et al. (2015, 2018) on plastic pollution in Africa. They show that 4.4 million metric tonnes of plastic waste were generated in the continent in 2010, projected to increase to 10.5 million tonnes by 2025, with Nigeria, Egypt and South Africa in waste generation by country at 5.96, 5.46 and 4.47 million tonnes per year in 2010 (Jambeck et al., 2015). They also show that the amount of mismanaged plastics due to inadequate disposal range from 23% to 85% in African countries where data is available, with most countries above 50%.

The slow progress in transitioning to a more circular plastic economy is associated with the complexity of moving a sector away from the long-ingrained take, make, use and dispose linear model to one that requires a complete paradigm shift. Many existing structures that enable the linear plastic economy model automatically pose a barrier to achieving this transition. Hence, enabling conditions need to be in place to scale up the solutions to achieve the desired transformation. The rest of this chapter discusses the challenges of transitioning to a circular plastic economy and the enabling conditions needed for success in Africa.

3 Barriers and enablers of a circular plastic economy in Africa

The interaction between the technical solutions for a circular plastic economy (discussed in Section 2), the barriers to achieving the circular plastic economy goals and the enabling conditions needed for success are illustrated using the lever diagram in Figure 2.2. The barriers hinder the scaleup of the solutions described in Section 2, making the circular plastic economy transition more challenging. Enablers, represented by the fulcrum, are structures that could make the transition to a circular plastic economy easier if in place.

3.1 Regulatory and institutional

The effective management of plastics is contingent on having a robust waste management regulatory and policy framework. A significant move towards a

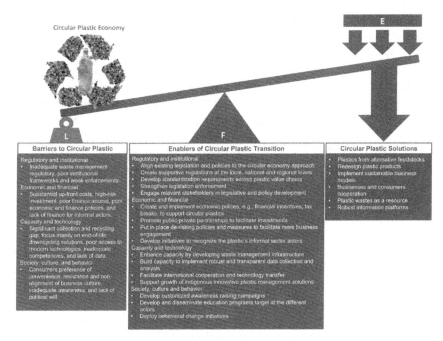

Circular Plastic Economy

E

L

F

Barriers to Circular Plastic	Enablers of Circular Plastic Transition	Circular Plastic Solutions
Regulatory and institutional • Inadequate waste management regulatory, poor institutional frameworks and weak enforcements Economic and financial • Substantial up-front costs, high-risk investment, poor finance access, poor economic and finance policies, and lack of finance for informal actors. Capacity and technology • Significant collection and recycling gap, focus mainly on end-of-life downcycling solutions, poor access to modern technologies, inadequate competencies, and lack of data. Society, culture, and behavior • Consumers preference of convenience, resistance and non-alignment of business culture, inadequate awareness, and lack of political will	Regulatory and institutional • Align existing legislation and policies to the circular economy approach • Create supportive regulations at the local, national and regional levels • Develop standardization requirements across plastic value chains • Strengthen legislation enforcement • Engage relevant stakeholders in legislative and policy development Economic and financial • Create and implement economic polices, e.g., financial incentives, tax breaks, to support circular plastics • Promote public-private partnerships to facilitate investments • Put in place de-risking policies and measures to facilitate more business engagement • Develop initiatives to recognize the plastic's informal sector actors Capacity and technology • Enhance capacity by developing waste management infrastructure • Build capacity to implement robust and transparent data collection and analysis • Facilitate international cooperation and technology transfer • Support growth of indigenous innovative plastic management solutions Society, culture and behavior • Develop customized awareness raising campaigns • Develop and disseminate education programs target at the different actors • Deploy behavioral change initiatives	• Plastics from alternative feedstocks • Redesign plastic products • Implement sustainable business models • Businesses and consumers cooperation • Plastic wastes as a resource • Robust information platforms

FIGURE 2.2   A lever diagram summarising the interactions between the available solutions (effort, E) for transitioning to a more circular plastic economy, the barriers to achieving the transition (the load, L) and the enablers (fulcrum, F) needed to change the plastic sector from a linear to a circular economy. The availability of circular plastic solutions without the required enablers will not be sufficient to achieve a circular plastic economy in Africa or elsewhere

better waste management regulatory framework has been observed in Africa recently. At least 50 out of 54 countries in Africa have developed policies or legislation addressing waste management (Attafuah-Wadee and Tilkanen, 2020). African countries now have the highest number of laws targeting plastic bags globally (Nyathi and Togo, 2020). However, it is still essential that policy and legislative frameworks are better tailored to achieve a circular plastic economy. Banning plastic bags is not enough to cause the needed shift. Specific legislations mandating circular plastic product design, reuse/recycling requirements and those facilitating sorting and collection of plastic wastes are still lacking but are essential for achieving a paradigm shift. For example, introducing extended producer responsibility (EPR) policies and legislation within the plastic value chain would shift plastic management responsibilities to producers; facilitate circular product designs; and encourage the collection, reuse and recycling. Sixteen African countries have introduced EPR-related policies on plastics and other products, including Ghana, Nigeria and South Africa (Ajani and Kunlere, 2019; Attafuah-Wadee and Tilkanen, 2020; Arp, 2021; Holland, 2021). It is essential that these efforts become continent-wide and address all common plastic types in the continent. Also, clear legislative guidance on using recycled plastic is lacking but needed. While significant discussions are ongoing on the use of recycled plastics in food packaging, there is no regulation, standard or guidance on this in African countries. Van Os and De Kock (2021) identified this lack of guidance as one of the reasons for the low plastic recycling rate in South Africa; it makes it difficult for plastic recyclers, businesses and entrepreneurs to be confident in investing in the sector. Furthermore, boosting the recovery of used plastics through business models such as product takeback will need to be supported by standards and policies that enhance the ability to monitor and track material flows, which are currently lacking in many countries.

Sometimes, existing legislation or policies could be obstructive, thus undermining a circular plastic economy. For example, trade policies that encourage the transfer of waste plastics from other countries to African nations are counterproductive (for instance, Ngcuka, 2021) and will make achieving a circular plastic economy more difficult. African countries, including Nigeria, Senegal, Morocco and South Africa, were destinations for waste plastics and scraps in 2018 (Pacini et al., 2021). Yet, the right trade laws and agreements can be leveraged to facilitate access to circular plastic products while preventing the import of uncompliant products (Ugorji and van der Ven, 2021). Also, regulatory and policy frameworks may be inadequate for a circular plastic economy in some countries because of their alignment with traditional manufacturing pathways, which are based on the continued extraction of non-renewable virgin resources.

Further, some countries' legislation focuses on the end-of-pipe minimisation of the adverse impacts of plastics rather than addressing product design or the complete plastic life cycle – i.e., production, use, distribution, trade and disposal (Excell, 2019). UNEP (2018) indicates that some African countries' legislation only focuses on some aspects of the life cycle rather than the whole. All African countries need to enact legislation that addresses the whole plastic life cycle. Countries that manufacture plastics need to ensure that policies and laws encourage the redesign of plastic products and production from renewable feedstocks, including bio-based sources. Countries that import plastics need to have policies and regulations that ensure that only circular plastic products are allowed.

Many of the current institutional frameworks are not adequately suited for a transition to a circular plastic economy. Circular plastic solutions require addressing material resources' consumption while considering the interlinkages with the economy, society and environment – requiring a whole-of-society approach. However, many African countries' public and private institutions often operate in silos and are not geared towards an integrated approach. For the circular plastic economy to succeed, the public and private sectors and civil society actors need to work together to stimulate innovation, develop appropriate solutions and mobilise resources and expertise towards a common goal. Further, a transition to a circular plastic economy will require all relevant departments in the public sectors, such as finance, natural resources, environment, mining, agriculture, energy, labour, education, etc., to cooperate beyond administrative silos towards a unified objective. But implementing the required organisational change can be difficult. Restructuring organisations to effectively address the needs of a circular economy can be expensive, risky and may induce resistance (Oghazi and Mostaghel, 2018).

The poor institutional framework results in the lack of effective implementation or enforcement of existing legislation, as has been noted in many African countries (for example, Chasse, 2018; Adebiyi-Abiola et al., 2019; Adam et al., 2020; Nyathi and Togo, 2020; Behuria, 2021). Of the 34 African nations that have banned or partially banned plastic bags, 16 are yet to introduce corresponding legislation for enforcement (Greenpeace Africa, 2020). Nyathi and Togo (2020) noted that implementing and enforcing plastic legislation in many African countries is usually weak due to inconsistencies. Stakeholder (especially the plastic industries) resistance was also pointed out as a reason for poor enforcement (Nyathi and Togo, 2020; Behuria, 2021), highlighting the need for a whole societal approach in developing and implementing legislation. Contributing to a lack of adequate implementation and enforcement is the fragmentation of regulatory systems in some countries where standards and responsibilities at the national or local levels are not clear, resulting in an inadequate legal system and poor accountability. IUCN (2020) noted this fragmentation as part of the challenges affecting the legal and institutional framework for marine plastic management in South Africa. However, the challenge of fragmented legal and institutional frameworks is common not just to Africa, as noted by other scholars (for example, Dauvergne, 2018; Nielsen et al., 2019; WWF and Dalberg, 2021).

To overcome the discussed barriers, the following legislative and institutional enablers should be in place:

- Align existing legislation and policies (e.g., waste management and natural resource management policies) to the circular economy approach. This could involve mandating the reuse and recycling of plastics, facilitating the sorting and collection of plastic wastes and ensuring that natural resource extraction policies follow environmental sustainability practices.
- Create supportive regulations at the local, national and regional levels that address the different types of plastics used predominantly in African countries (beyond plastic bags). And these legislations should address the whole life cycle of these plastics – design, production, importation, use, distribution, trade, reuse, recycling and disposal.
- In line with the above, develop standardisation requirements across the plastic product value chain at the national and regional levels to promote circularity and develop EPR legislation, plastic takeback laws or other similar legislation to ensure that plastic manufacturers take responsibility for end-of-life management of their products.

- Address the challenge of poor legislation enforcement by implementing enforcement strategies and mechanisms. This could include addressing fragmentation; strengthening institutions (e.g., increasing personnel and building their capacity); giving greater power to environmental authorities; revamping and ensuring consistent implementation of environmental permits, licenses and certificates; and applying appropriate penalties for non-compliance.
- Engage relevant stakeholders, including the plastic industry and importers, recyclers and civil society, in developing and implementing relevant plastic legislations and commitments. This could be in the form of the national plans for plastic management like the SA Plastic Pact developed by the South African government in collaboration with stakeholders (see: www.saplasticspact.org.za/). Such action plans, roadmaps and standards are essential prerequisites for attracting funding for circular solutions (Schroder and Raes, 2021).

3.2 Economic and financial

Circular plastic solutions such as production from alternative feedstock, waste collection and sorting, upcycling plastic waste into new valuable products or establishing industrial symbiosis or urban–industrial symbiosis will incur substantial up-front costs, including the cost of installing new infrastructure, retrofitting existing production systems, building new distribution and logistical arrangements and retraining staff (Ambrose 2019; Preston et al., 2019; Davies et al., 2020). A SWOT analysis of the feasibility of establishing a plastic recycling facility in East Africa highlighted high cost as one of the main weaknesses of the project (Davies et al., 2020). This high cost is a deterrent for many investors who perceive the waste management sector as a high-risk investment in Africa (UNEP 2018). More so, a lack of effective financing models, inadequate institutional frameworks and poor governance of public resources was noted as a major contribution to insufficient finance and investment in waste management solutions in many African countries (Godfrey et al., 2019).

Furthermore, many plastic management facilities in Africa lack access to adequate finance, particularly small- or medium-scale enterprises that may have limited creditworthiness, and collateral, and could be risk-averse. Also, the long lead time to break even for plastic management facilities is a barrier. For example, bio-based production or upcycling installations may take time to deliver higher yields and revenues, which might not align with investors' interests or conditions. It is, however, heartwarming to note a first-of-its-kind investment in which Dow (a foreign material science company) and other investors are providing funds to an African start-up company, Mr. Green Africa, focused on accelerating the circular plastic economy in the continent (Magoum, 2022). This type of financing model needs to be studied and explored further in other African countries.

An important factor that makes plastic waste management, e.g., recycling, less viable is the apparent lower price of virgin plastic feedstocks than recycled polymers (Hopewell et al., 2009). Virgin feedstock seems cheaper partly due to fossil-fuel subsidies. Cabernard et al. (2022) noted a significant increase in South Africa's plastic-related greenhouse gas emissions, with 95% attributed to domestic coal use; yet the government provides a substantial subsidy to the country's coal production (Doukas and Roberts, 2019; Pant et al., 2020). Another reason is that economic policies and investment and financing decisions do not incorporate the non-monetised negative externalities (e.g., adverse environmental and human health effects) associated with plastic production from virgin feedstocks. Deloitte (2019) found that the economic impact of plastic pollution in the African region ranges between $33 million and $161 million (i.e., $25–$69 million in cleanup costs and $8–$92 million in lost revenue). WWF and Dalberg (2021) found that plastic production in 2019 had a minimum lifetime cost of between $43 and $78 billion when the damage to livelihoods and other economic sectors and cleanup costs and adverse effects on human health are considered. Globally, they found that the cost of plastic production in 2019 to society is at least $3.7 trillion. Because these externalities are not accounted for in many economic policies (not only in Africa), the linear plastic consumption model appears to be profitable and adequate. These economic policy deficiencies make circular plastic solutions less attractive to businesses – which often depend on supportive policies to help cushion the challenges of shifting to different business models (Preston, 2012; Preston et al., 2019).

Also, a thriving circular economy will depend significantly on recognising and integrating the informal sector and smallholder businesses into the formal economy. The informal sector and smallholder businesses form a significant portion of the world's economy and play a critical role in plastic waste management in Africa (for example, Oteng-Ababio, 2012; Plastics SA, 2019a; Gall et al., 2020; GAIA, 2021; IUCN, 2021b). Unfortunately, these important actors are not adequately recognised in many countries. Current institutional arrangements disconnect them from the formal economy, thus inhibiting their effective contribution to the circular plastic economy transition (WBCSD, 2016; Yeoh, 2020). This situation also makes it more challenging to reach them with policies or bring in new ideas and technologies (Preston et al., 2019). Yet, this underappreciated sector plays a significant role in plastic recycling globally – more than half of the plastics (59%) recycled globally (i.e., 27 million metric tonnes) in 2016 were collected by the informal sector (PEW, 2020). In South Africa, the more than 58,000 informal plastic waste pickers were responsible for 70% of recycled plastics (Plastics SA, 2019b). Training and organising small businesses (for example, into cooperatives or associations) and supporting their recognition by government and financial institutions could help strengthen their competitiveness and improve their access to finance for circular solutions (Medina, 2005; Buch et al., 2021). Kasinja and Tilley (2018) indicated that the organisation of plastic and metal waste pickers into waste management cooperatives in Malawi could enhance their activities while also providing other social and economic benefits. GAIA (2021) highlighted examples from South Africa, Ghana, Tanzania, Kenya, Morocco and Zambia, where organising waste pickers into cooperatives and associations led to more access to finance while increasing their contributions to achieving a more circular economy and improving their socioeconomic status.

The following are, therefore, essential to address the economic and financial barriers to a circular plastic economy:

- Implementing economic policies that support circular plastic solutions, for example, tax breaks and subsidies. The cost of using products made from post-consumer plastic wastes (for example, bricks made from recycled plastics) could be reduced through subsidies or tax credits. Credits and subsidies could also be extended to businesses involved in plastic pollution management. Concurrently, higher taxes for plastics containing a high amount of virgin feedstocks and removal of subsidies for plastics feedstocks such as coal (in recognition of their negative externalities) could also be instituted to enhance the competitiveness of circular plastic products.

- Promote public–private partnerships (PPPs) to facilitate investments in circular plastic and waste management solutions. PPP could help access additional funding and finance, provide technical expertise and innovation that are usually stronger in the private sector and help develop win-win solutions for all stakeholders. An example is a PPP between the Rwanda Environment Management Authority and the Private Sector Federation, which aims to raise more than $700,000 to address single-use plastics. The private sector finance would support the collection, disposal and recycling of plastics, while the public sector would ensure technical expertise and public awareness (REMA, 2020; YeniSafak, 2021).

- Putting in place de-risking policies and measures to enable small and medium businesses and large businesses to get more involved in circular plastic solutions. This is important to manage the high risk, high infrastructure cost and long lead time associated with circular economy business models and make these important actors and their projects more bankable (Schroder and Raes, 2021). Tax exemptions, subsidies and other fiscal incentives could be great de-risking measures and instruments such as blended finance and investment guarantees (Schroder and Raes, 2021). Also, capacity-building initiatives, green investment policies and initiatives such as regional and national green bonds or creating dedicated financial instruments for the circular plastic economy are ways to make it easier for small and medium businesses to cushion the initial cost of circular plastic investments (European Commission, 2019; Schroder and Raes, 2021).

- Develop and implement initiatives to help recognise the vital role of the informal sector in the circular plastic economy. As noted earlier, creating cooperatives or associations that bring the informal sector together can strengthen their contributions to a circular plastic economy and provide other socioeconomic benefits. Doing this can also enhance their creditworthiness, thus providing better access to finance. Training and capacity-building initiatives can also enhance their contribution to a circular plastic economy.

3.3   *Capacity and technology*

The current capacity for managing the volumes of plastics produced/imported, used and disposed of in Africa is insufficient. EY (2020) noted a significant plastic collection and recycling gap in African cities, with interventions limited to fragmented solutions rather than a genuinely circular approach. They noted an 83-kt gap in Accra and Kampala and a 480-kt gap in Lagos, Nairobi and Addis Ababa. This capacity gap is connected to these cities' lack of adequate waste management infrastructure. The fluid and fractured nature of Africa's plastic recycling industry, characterised by frequent market entry and exit by players, is also a significant contributor to the gap (Holland, 2021). Addressing these capacity gaps will require innovative plastic management solutions.

Today's deep-rooted end-of-life technologies predominant in African countries are inadequate as most plastic wastes are mainly reprocessed into products or materials of lower value. Of the 14% of plastic waste collected globally in 2013, only 2% were recycled into products of the same or similar quality globally, 8% were downcycled, and the rest were lost (WEF 2016). While specific data comparing upcycling and downcycling is scarce for Africa, it is clear that plastic management is mainly dominated by downcycling plastics into products of lower value (WWF, 2018). This is due to the poor design of many plastic products, which makes them unsuitable for continued recycling, highlighting the need for circular designs. Another reason is the dominance of mechanical recycling in sub-Saharan Africa, with minimal application of chemical recycling solutions (EY, 2020). A successful transition to a circular plastic economy in the continent will require developing technological and innovative solutions that are fit for each country's unique situation to address these challenges and promote a new way of producing, using and managing plastics throughout their life cycle across the value chain.

Smart infrastructure and digital technologies are being developed and implemented for managing plastic across its life cycle, including for manufacture, waste sorting and data collection and tracking of products. For example, three-dimensional (3D) printing solutions have been developed for upcycling waste plastics (e.g., Gaikwad et al., 2018; Mikula et al., 2021), and artificial intelligence, blockchain and machine learning are being deployed for plastic waste sorting (Chidepatil et al., 2019; Gupta et al., 2019; Ioannou and Petrova, 2019; University of Sydney, 2021). Blockchain has also been deployed to enhance plastic traceability and facilitate recycling while providing socioeconomic benefits for plastic recyclers (Katz, 2019; Taylor et al., 2020). But there remain significant capacity and technology access barriers to deploying or scaling up these high-tech solutions or making them widely accessible in low-income nations in Africa. Solutions requiring substantial energy use, internet access and other information technology facilities may be more difficult to deploy because of a lack of the enabling infrastructure and, sometimes, needed expertise. Many African countries still need to develop basic waste management infrastructure and may not be able to invest in these advanced technologies.

Furthermore, emerging business models for a circular plastic economy, such as product redesign, reuse, refill, reverse logistics, product takeback and urban and industrial symbiosis, would require new expertise across organisation structure, within government departments and in the plastic supply chains. Developing these new competencies may be challenging and delay the quick adoption of circular solutions. Inadequate knowledge among businesses and the public sector, the lack of comprehensive training and skill development, insufficient knowledge dissemination among stakeholders, and the potential higher costs of business restructuring and capacity building are drags to the effective adoption of new technologies, solutions and business models that will be typical to a circular plastic economy.

Underlying the implementation of circular plastic solutions is the availability of relevant data on resource flow, but this is also a significant challenge in Africa, as have been noted, for example, by Jambeck et al. (2018) and Andriamahefazafy and Failler (2022). The lack of data and, importantly, the infrastructure and capacity to collect them make it difficult to develop adequate strategies and supporting policies for a circular plastic economy. It also makes it difficult to assess the effectiveness of implemented solutions and hinders collaboration across businesses, the public sector and other stakeholders.

However, some progress is being achieved in the continent concerning bridging the plastic management capacity gap, deploying new technologies and solutions and addressing the data gaps spurred by the recent engagement of start-ups and entrepreneurs in plastic management (EY, 2020). Some examples are highlighted in Footprints Africa (2021), UNDP (2019), Adebiyi-Abiola et al. (2019) and Oyinlola et al. (2022), including web and mobile platforms facilitating plastic waste collection, upcycling of plastics into valuable products, refill business models for plastic containers and plastic recycling. 3D printing has also been used to convert plastic waste to low-cost agricultural tools for African farmers through an international partnership project involving Kenya, Rwanda, Nigeria and Loughborough University in the United Kingdom (Magoum, 2021). Furthermore, initiatives such as the African Plastics Recycling Alliance, which aims to significantly transform plastic recycling infrastructure in the continent (IISD, 2019), could help bridge the capacity and technological gaps.

To build on these ongoing efforts, the following actions could be taken:

- Enhance capacity to manage plastics by developing waste management infrastructure (in rural and urban areas) and human resources needs (technical and managerial) on best practices of plastic management. The capacity to develop policies and legislation and monitor progress towards a circular plastic economy should also be developed.
- Address data needs by building capacity to implement robust and transparent data collection and analysis. This will provide needed information on plastic resource flows and facilitate better collaboration to implement circular plastic interventions.

- Facilitate international cooperation and technology transfer between countries within and outside the continent to promote learning and capacity building. This could be through south–south collaboration that brings public and private sector actors, academics and researchers, civil society and other stakeholders together to collaborate and exchange best practices on plastic management. Other options for cooperation could be through international initiatives such as the World Economic Forum's Global Plastic Action Partnership which Ghana and Nigeria are part of (WEF, 2021). This supports the two countries in developing a roadmap and common approach to plastic management.
- Facilitate the growth of indigenous innovative plastic management solutions and entrepreneurs through incubator programmes, networking opportunities, innovation prizes that can foster experimentation and the development of technologies that address the unique concerns of the continent and accord with national contexts, socioeconomic circumstances and cultural realities. Successful examples should be disseminated as best practices. Entrepreneurial and business growth could also be enhanced through product standards and specifications requiring recycled plastics in products, thereby increasing the demand for circular plastic products.

3.4 Society, consumer, business and government culture and behaviour

Societal and cultural factors relate to how much consumers, businesses and government institutions are aware of, embrace, and are willing to implement a circular plastic economy. With consumers, preference for convenience over socially and/or environmentally beneficial practices remains a significant challenge and is a key factor that has fuelled the use of single-use plastics in Africa (for example, Verghese et al., 2008; Adane and Muleta, 2011; O'Brien and Thondhlana, 2019). It is much easier to implement circular solutions and initiate positive behavioural change when the concept and value of environmental and socioeconomic benefits of a circular plastic economy are recognised and understood by consumers (Moss, 2021).

While awareness of the negative effect of plastics is increasing, at least among many urban dwellers, it is yet to translate into significant change because changing people's behaviour is complex. Even acquiring a higher education does not necessarily translate to action, as noted in South Africa by O'Brien and Thondhlana (2019) where high spending consumers were willing to spend more to use plastic bags. This highlights a need for more educational initiatives specifically tailored to promote citizen awareness and behavioural nudge interventions to help consumers act appropriately. It should also be noted that consumers also stated the lack of alternatives or substitutes for plastics as a reason why some have been unable to change (Verghese et al., 2008; Adebiyi-Abiola et al., 2019; Adam et al., 2020). Language barrier is also critical, especially among rural African dwellers, as circular principles need to be broken down to local context.

Resistance and non-alignment of businesses and company culture with the principles of a circular plastic economy could be a barrier among businesses. While a circular plastic economy will require significant business-to-business cooperation along the supply chain, many organisations are either unwilling to cooperate or not designed for these types of collaboration. The small size and organisational structure (or a lack of it) of many actors along the plastic supply chain in Africa could precipitate a culture that makes it challenging to collaborate with other businesses. The need for a long-term business perspective and the required significant change to business models in a circular plastic economy could also lead to resistance among business leaders whose short-term interests may dominate decision-making and prefer the status quo (Houston et al., 2018; Behuria, 2021). Inadequate awareness and communication culture within organisations and between businesses involved in the plastic supply chain means that many may not understand the benefits, making it difficult to support new solutions, change business models or collaborate with others.

On the government side, a lack of political will to promote the circular economy and to lead by example could also be an obstacle. For consumers, businesses and other stakeholders to take the transition to a circular economy more seriously, it is essential for the linear plastic use culture within governmental institutions to change, for example, in procurement. Eliminating single-use plastics and other non-circular plastic products in government-related buildings and events and promoting sustainable behaviour among government workers can set a good example for businesses and influence consumers' choices (Environment Georgia, 2021). Governments also have a significant role in promoting the right behaviour among companies and individuals through policies, regulations, incentives, standards and awareness-raising (Sections 3.1 and 3.2).

To enable the transition to a circular plastic economy, the following actions and nudges need to be put in place to help overcome sociocultural barriers:

- Develop customised awareness-raising campaigns tailored towards the desired change expected from consumers, businesses, decision-makers and institutions. It could be anchored as part of a broader sustainable development objective and include explicit language and messages and the expected outcomes for each constituency. Awareness campaigns should employ relevant media channels that make it easy to reach different actors, such as social media for youths and the middle class; radio and community meetings for rural dwellers; print, digital and social media for urban dwellers; etc. Being specific on what needs to be done (e.g., separating plastic at the source or choosing refillable products) and adding humour to awareness initiatives make them memorable and can promote adoption (Kelleher, 2021; Moss, 2021).
- Develop and disseminate educational programmes targeted at the different actors in the plastic value chain. Circular economy and sustainability concepts and principles should be incorporated into the education curriculum at all levels and

training programmes in the public and private sectors. Educational platforms for sharing resources and best practices should be developed for businesses and government institutions. Best practices on green procurement should be taught in government institutions. Educational programmes should also target rural dwellers using appropriate language and media such as community radio and television.

- Deploy initiatives to encourage a change in behaviour among consumers, businesses and government institutions. Behaviour change can be promoted through incentives or disincentives for consumers and businesses (Jia et al., 2019; Metternicht et al., 2020; Kelleher, 2021). For example, charges on single-use plastics, donations by supermarkets to charity for each plastic avoided by customers (Adeyanju et al., 2021) or subsidies on circular plastics to promote change in business behaviour. Rules and regulations to require or encourage desired behaviour (e.g., bans of non-circular plastics or charges for the use of certain plastic packaging) could also be instituted. Providing information about the desired behaviour, for example, through awareness campaigns (see above), can also help nudge individuals, businesses and government in the right direction, as well as influence through peers and social groups (Rapada et al., 2021) and the availability of alternatives that make decision-making easier.

4 Plastics management, systems thinking and sustainable developmental priorities

Creating the enabling environment for a successful transition to a circular plastic economy in Africa (and elsewhere) would require addressing the different components that form the plastic resource system. Using a modification of the conceptual model developed by Iacovidou et al. (2021), the plastic resource system comprises the material flow subsystem (extraction, processing, production, importation, consumption, reuse, recycling and disposal of plastic resources) and actors' subsystem (manufacturers, businesses, investors, retailers, waste management industry, government and consumers) interacting with each other (Figure 2.3). Each actor within the plastic resource system plays different roles based on their values and goals. For example, at the basic level, the manufacturer's goal is to make plastic products using available natural resources, while the investors' goal is to maximise the return on their investments. Consumers generally want products that are convenient and meet their needs, while businesses and retailers seek to meet consumers' needs and make profits in the process. All actors within the system will act to promote their objectives.

But the plastic resource system is further embedded into a broader system comprising the environment and associated ecosystem services; technologies and innovations; governance, regulatory, policy and institutional frameworks; economic, financial and market influences; and human and societal needs and behaviour. These broader system components constantly interact with each other

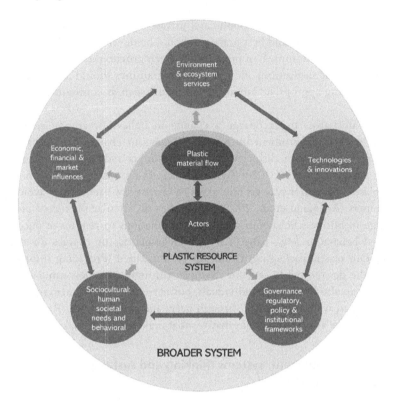

FIGURE 2.3   The plastic resource system comprises resource flow and actors embedded in a broader system. The wider system interacts with the resource system, affecting the entire system's behaviour directly or indirectly

and with the plastic resource subsystem (Figure 2.3). For example, the environment is the source of the natural resources used in making plastics (e.g., biomaterials and fossil fuels). The type of technology and innovation applied in extracting, processing and managing the natural resources (to make plastics) will influence how the production of plastics impacts the environment and associated ecosystem services. Hence, non-circular plastic production is associated with significant environmental impacts, as described in Section 1, while circular plastics would have less impact.

Policies, regulations, institutional frameworks and the prevailing economic and market situation would influence how plastics are made and their flow within the resource system. Further, prevailing economic realities and broader policy priorities can significantly influence societal behaviour and determine whether a circular plastic economy can be achieved. An example of how socioeconomic reality affects human behaviour towards plastic management is in Nigeria and Cameroon, which faces the challenge of plastic pollution due to inappropriate use and disposal of water sachets – because of a lack of access to drinking water and the high cost of alternative packaged water. Nigeria is most likely the highest consumer of water sachet in Africa. But consumers are unable to reject the water sachet due to a lack of public water supply, thus resulting in a continuous plastic pollution cycle (Adebiyi-Abiola et al., 2019; Kobo, 2021). Another example is in Cote D'Ivoire and Ghana, where it was challenging to implement a ban on plastic bags because plastic bag production employs a substantial number of citizens (up to 1 million citizens in Cote D'Ivoire) (Kobo, 2021). These highlight the interlinkages between the plastic resource system and societal needs and the importance of connecting circular plastic solutions with immediate socioeconomic priorities. Sustainable development priorities such as job creation, economic diversification, better health, sanitation, food security, equality and poverty reduction are critical components of the broader system on which the transition to a circular plastic economy can be anchored to gain traction, attract investments and be successful.

Therefore, to effectively manage plastics throughout their life cycle, solutions should be based on the systems thinking approach. The approach ensures that the complex interactions between the various components that form the plastic resource subsystem and those of the broader system are considered to achieve desired outcomes across all three dimensions of sustainable development (environment, social and economic). Understanding these interactions will help comprehend the root causes and support the development of appropriate interventions that will not lead to adverse effects on other system components. For example, in developing an action plan for plastic management, an understanding of the interactions within the plastic resources system and with the broader socioeconomic system in a country or city can help guide in determining which of the enablers discussed in Section 3 should be prioritised at the national or city levels. The outputs from such analysis are expected to differ in each African country. But by applying the systems thinking approach, it is possible to develop solutions that consider the specific national or local contexts, prevailing economic and social circumstances and cultural realities.

Governments have a critical role in achieving a circular plastic economy because of their significant power to put in place the required enablers. It can develop regulations and policies, facilitate technology access and implement capacity-building initiatives. But government efforts must incorporate all relevant stakeholders (an essential aspect of the systems thinking approach) and align with a vision of sustainable development across all economic sectors. That way, it can ensure that solutions to plastic pollution address root causes, achieve the targeted objectives and do not have unintended negative consequences on other environmental, societal and economic priorities.

References

Adam, I., Walker, T.R., Bezerra, J.C., Clayton, A. (2020) Policies to reduce single-use plastic marine pollution in West Africa. *Marine Policy*, 116, https://doi.org/10.1016/j.marpol.2020.103928

Adane, L., Muleta, D. (2011) Survey on the usage of plastic bags, their disposal and adverse impacts on environment: A case study in Jimma City, Southwestern Ethiopia. *Journal of Toxicology and Environmental Health Sciences*, 3, 234–248.

Adebiyi-Abiola, B., Assefa, S., Sheikh, K., Garcia, J.M. (2019) Cleaning up plastic pollution in Africa. *Science*, 365, 1249–1251. https://doi.org/10.1126/science.aax3539

Adeyanju, G.C., Augustine, T.M., Volkmann, S., Oyebamiji, U.A., Ran, S., Osobajo, O.A., Otitoju, A. (2021) Effectiveness of intervention on behaviour change against use of non-biodegradable plastic bags: a systematic review. *Discover Sustainability*, 2, 13. https://doi.org/10.1007/s43621-021-00015-0

Ahmetli, G., Kocaman, S., Ozaytekin, I., Bozkurt, P. (2013) Epoxy composites based on inexpensive char filler obtained from plastic waste and natural resources. *Polymer Composite*, 34. https://doi.org/10.1002/pc.22452

Ajani, I.A., Kunlere, I.O. (2019) Implementation of the extended producer responsibility (EPR) policy in Nigeria: Towards sustainable business practice. *Nigerian Journal of Environment and Health*, 2, 44–56.

Alberghini, M., Hong, S., Lozano, L.M., Korolovych, V., Huang, Y., Signorato, F., Zandavi, S.H., Fucetola, C., Uluturk, I., Tolstorukov, M.Y., Chen, G., Asinari, P., Osgood III, R.M., Fasano, M., Boriskina, S.V. (2021) Sustainable polyethylene fabrics with engineered moisture transport for passive cooling. *Nat Sustain*, 4, 715–724. https://doi.org/10.1038/s41893-021-00688-5

Ambrose, J. (2019) War on plastic waste faces setback as cost of recycled material soars. *The Guardian*. Available at: www.theguardian.com/environment/2019/oct/13/war-on-plastic-waste-faces-setback-as-cost-of-recycled-material-soars

Andriamahefazafy, M., Failler, P. (2022) Towards a circular economy for African islands: an analysis of existing baselines and strategies. *Circular Economy and Sustainability*, 2, 47–69. https://doi.org/10.1007/s43615-021-00059-4

Appiah, J.K., Berko-Boateng, V.N., Tagbor, T.A. (2017) Use of waste plastic materials for road construction in Ghana. *Case Studies in Construction Materials*, 6, 1–7. http://dx.doi.org/10.1016/j.cscm.2016.11.001

Arp, R. (2021) Extended producer responsibility for plastic packaging in South Africa: A synthesis report on policy recommendations. WWF South Africa, Cape Town, South Africa.

Attafuah-Wadee, K., Tilkanen, J. (2020) Policy approaches for accelerating the circular economy in Africa. Circular Economy. *Earth*. Available at: https://circulareconomy.earth/publications/accelerating-the-circular-economy-transition-in-africa-policy-challenges-and-opportunities

Ayeleru, O.O., Dlova, S., Akinribide, O.J., Ntuli, F., Kupolati, W.K., Marina, P.F., Blencowe, A., Olubambi, P.A. (2020) Challenges of plastic waste generation and management in sub-Saharan Africa: A review. *Waste Management*, 110, 24–42. https://doi.org/10.1016/j.wasman.2020.04.017

Babayemi, J.O., Nnorom, I.C., Osibanjo, O., Weber, R. (2019) Ensuring sustainability in plastics use in Africa: Consumption, waste generation, and projections. *Environmental Sciences Europe*, 31, 60. https://doi.org/10.1186/s12302-019-0254-5

Barra, R., Leonard, S.A. (2018) Plastics and the circular economy. A STAP document. Scientific and Technical Advisory Panel to the Global Environment Facility, Washington, DC. Available at: https://stapgef.org/sites/default/files/2020-02/PLASTICS%20for%20posting.pdf?null=

Beach, E., Weeks, B.R., Stern, R., Anastas, P. (2013) Plastics additives and green chemistry. *Pure and Applied Chemistry*, 85, 1611–1624. http://dx.doi.org/10.1351/PAC-CON-12-08-08

Behuria, P. (2021) Ban the (plastic) bag? Explaining variation in the implementation of plastic bag bans in Rwanda, Kenya and Uganda. *Environment and Planning C: Politics and Space*, 39, 1791–1808. https://doi.org/10.1177%2F2399654421994836

Buch, R., Marseille, A., Williams, M., Aggarwal, R., Sharma, A. (2021) From waste pickers to producers: An inclusive circular economy solution through development of cooperatives in waste management. *Sustainability*, 13, 8925. https://doi.org/10.3390/su13168925

Cabernard, L., Pfister, S., Oberschelp, C., Hellweg, S. (2022) Growing environmental footprint of plastics driven by coal combustion. *Nature Sustainability*, 5, 139–148. https://doi.org/10.1038/s41893-021-00807-2

Chasse, C. (2018) Evaluation of legal strategies for the reduction of plastic bag consumption. Master's thesis, Harvard Extension School. Available at: https://dash.harvard.edu/bitstream/handle/1/42004017/CHASSE-DOCUMENT-2018.pdf

Chidepatil, A., Bindra, P., Kulkarni, D., Qazi, M., Kshirsagar, M., Sankaran, K. (2019) From trash to cash: How blockchain and multi-sensor-driven artificial intelligence can transform circular economy of plastic waste? *Administrative Science*, 10, 23. https://doi.org/10.3390/admsci10020023

Dauvergne, P. (2018) The power of environmental norms: Marine plastic pollution and the politics of microbeads. *Environmental Politics*, 27. https://doi.org/10.1080/09644016.2018.1449090

Davies, B., D'Andrea, M., Jarvi, L., Sporney, M., Urich, M. (2020) DU social good project – Africa recycling feasibility. University of Denver EMBA Cohort #73. Available at: https://static1.squarespace.com/static/5be066eca9e028fe088be707/t/5f33fca10726ac1603cb5fa9/1597242530370/Aftrica+Recycling+Plant+Feasibility+-+University+of+Denver+EMBA.pdf

Deloitte. (2019) The price tag of plastic pollution. An economic assessment of river plastic. *Deloitte*. Available at: www2.deloitte.com/content/dam/Deloitte/nl/Documents/strategy-analytics-and-ma/deloitte-nl-strategy-analytics-and-ma-the-price-tag-of-plastic-pollution.pdf

Doukas, A., Roberts, L. (2019) G20 coal subsidies: South Africa. Overseas Development Institute. Available at: https://odi.org/en/publications/g20-coal-subsidies-south-africa/

Environment Georgia. (2021) Clarkston becomes third municipality to phase out single use plastics in government buildings. *Environment Georgia News Release*. Available at: https://environmentgeorgiacenter.org/news/gae/clarkston-becomes-third-municipality-phase-out-single-use-plastics-government-buildings

European Commission. (2019) Accelerating the transition to the circular economy – Improving access to finance for circular economy projects. European Commission, Brussel, Belgium.

Excell, C. (2019) 127 Countries now regulate plastic bags. Why aren't we seeing less pollution? World Resource Institute. Available at: www.wri.org/insights/127-countries-now-regulate-plastic-bags-why-arent-we-seeing-less-pollution

EY. (2020) Mapping the integrated supply chain for plastics in Africa. Recycling plastic waste to create jobs. Ernst & Young Global Limited. Available at: www.transform.global/news/mapping-the-integrated-supply-chain-for-plastics-in-africa/

Footprints Africa. (2021) The circular economy: Our journey in Africa so far. Footprints Africa. Available at: https://irp-cdn.multiscreensite.com/40a0e554/files/uploaded/CEcasereport_Footprints.pdf

GAIA. (2021) Strengthening waste picker organizing in Africa. GAIA. Available at: www.no-burn.org/wp-content/uploads/2021/12/Strengthening-Waste-Picker-Organising-in-Africa.pdf

Gaikwad, V., Ghose, A., Cholake, S., Rawal, A., Iwato, M., Sahajwalla, V. (2018) Transformation of E-waste plastics into sustainable filaments for 3D printing. *ACS Sustainable Chemistry & Engineering*, 6, 14432–14440. https://doi.org/10.1021/acssuschemeng.8b03105

Gall, M., Wiener, M., de Oliveira, C.C., Lang, R.W., Hansen, E.G. (2020) Building a circular plastics economy with informal waste pickers: Recyclate quality, business model, and societal impacts. *Resources, Conservation and Recycling*, 156, 104685. https://doi.org/10.1016/j.resconrec.2020.104685

Godfrey, L., Ahmed, M.T., Gebremedhin, K.G., Katima, J.H.Y., Oelofse, S., Osibanjo, O., Richter, U.H., Yonli, A.H. (2019) Solid waste management in Africa: Governance failure or development opportunity? *Regional Development in Africa*. https://doi.org/10.5772/intechopen.86974

Greenpeace Africa. (2020) 34 Plastic bans in Africa. A reality check. Greenpeace. Available at: www.greenpeace.org/africa/en/blogs/11156/34-plastic-bans-in-africa/

Gu, Y., Tamura, M., Nakagawa, Y., Nakao, K., Suzuki, K., Tomishige, K. (2021) Direct synthesis of polycarbonate diols from atmospheric flow CO_2 and diols without using dehydrating agents, *Green Chemistry*, 23, 5786. https://doi.org/10.1039/D1GC01172C

Gupta, P.V., Shree, V., Hiremath, L., Rajendran, R. (2019) The use of modern technology in smart waste management and recycling: Artificial intelligence and machine learning. In: Kumar, R., Wiil, U. (eds) *Recent Advances in Computational Intelligence, Studies in Computational Intelligence*, 823. Springer. https://doi.org/10.1007/978-3-030-12500-4_11

Guzman, A.D.M., Munno, M.G.T. (2015) Design of a brick with sound absorption properties based on plastic waste and sawdust. *IEEE Access*, 3, 1260–1271. https://doi.org/10.1109/ACCESS.2015.2461536

Hatti-Kaul, R., Nilsson, L.J., Zhang, B., Rehnberg, N., Lundmark, S. (2020) Designing biobased recyclable polymers for plastics. *Trends in Biotechnology*, 38, 50–67. https://doi.org/10.1016/j.tibtech.2019.04.011

Holland, C.P. (2021) South Africa leading the African circular plastics industry. *Recycling Magazine*. Available at: www.recycling-magazine.com/2021/09/03/south-africa-leading-the-african-circular-plastics-industry/

Hopewell, J., Dvorak, R., Kosior, E. (2009) Plastics recycling: Challenges and opportunities. *Philosophical Transactions of the Royal Society B: Biological Sciences*, 27, 2115–226. https://dx.doi.org/10.1098%2Frstb.2008.0311

Houston, J., Casazza, E., Briguglio, M., Spiteri, J. (2018) Enablers and barriers to a circular economy. Stakeholder Views Report. Available at: www.r2piproject.eu/wp-content/uploads/2018/08/R2pi-stakeholders-report-sept-2018.pdf

Iacovidou, E., Hahladakis, J.N., Purnell, P. (2021) A systems thinking approach to understanding the challenges of achieving the circular economy. *Environmental Science and Pollution Research*, 28, 24785–24806. doi.org/10.1007/s11356-020-11725-9

IEF. (2021) From mattresses to sunglasses: Turning CO_2 into plastics. *International Energy Forum*. Available at: www.ief.org/news/from-mattresses-to-sunglasses-turning-co2-into-plastics

IISD. (2019) Companies launch African plastics recycling alliance. *IISD Knowledge Hub*. Available at: https://sdg.iisd.org/news/companies-launch-african-plastics-recycling-alliance/

Ioannou, L., Petrova, M. (2019). America is drowning in garbage. Now robots are being put on duty to help solve the recycling crisis. *CNBC*. Available at: www.cnbc.com/2019/07/26/meet-the-robots-being-used-to-help-solve-americas-recycling-crisis.html

IUCN. (2020) The legal, policy and institutional frameworks governing marine plastics in South Africa. Exchange of perspectives to define priorities. *IUCN Environmental Law Center*. Available at: www.iucn.org/sites/dev/files/content/documents/webinar_report_south_africa_05112020.pdf

IUCN. (2021a) Circular economy projects in Eastern and Southern Africa: Changing lives and reducing the flow of plastic waste into the marine environment. *International Union for Conservation of Nature*. Available at: www.iucn.org/news/eastern-and-southern-africa/202102/circular-economy-projects-eastern-and-southern-africa-changing-lives-and-reducing-flow-plastic-waste-marine-environment

IUCN. (2021b) Waste pickers role in plastic pollution reduction: The ones we cannot leave behind. *International Union for Conservation of Nature*. Available at: www.iucn.org/news/environmental-law/202104/waste-pickers-role-plastic-pollution-reduction-ones-we-cannot-leave-behind

Jambeck, J., Hardesty, B.D., Brooks, A.L., Friend, T., Teleki, K., Fabres, J., Beaudoin, Y., Bamba, A., Francis, J., Ribbink, A.J., Baleta, T., Bouwman, H., Knox, J., Wilcox, C. (2018) Challenges and emerging solutions to the land-based plastic waste issue in Africa. *Marine Policy*, 96, 256–263. https://doi.org/10.1016/j.marpol.2017.10.041

Jambeck, J. R., Geyer, R., Wilcox, C., Siegler, T. R., Perryman, M., Andrady, A., Narayan, R., Law, K. L. (2015) Plastic waste inputs from land into the ocean. *Science*, 347, 768–771. https://doi.org/10.1126/science.1260352

Jia, L., Evans, S., van der Linden, S. (2019) Motivating actions to mitigate plastic pollution. *Nature Communications*, 10, 4582. https://doi.org/10.1038/s41467-019-12666-9

Kasinja, C., Tilley, E. (2018) Formalization of informal waste pickers' cooperatives in Blantyre, Malawi: A feasibility assessment. *Sustainability*, 10, 1149. https://doi.org/10.3390/su10041149

Katz, D. (2019) Plastic Bank: Launching Social Plastic® revolution. Field Actions Science Reports [Online], 19, http://journals.openedition.org/factsreports/5478

Kelleher, K. (2021) Prevention, reduction and control of Marine Plastic Pollution in African and Indian Ocean Developing Island States. Southwest Indian Ocean Fisheries Project No. 2. World Bank/Indian Ocean Commission.

Khan, I.M., Kabir, S., Alhussain, M.A., Almansoor, F.F. (2016) Asphalt design using recycled plastic and crumb-rubber waste for sustainable pavement construction. *Procedia Engineering*, 145, 1557–1564. https://doi.org/10.1016/j.proeng.2016.04.196

Khoonkari, M., Haghighi, A.H., Sefidbakht, Y., Shekoohi, K., Ghaderian, A. (2015) Chemical recycling of PET wastes with different catalysts. *International Journal of Polymer Science*, 2015, 124524, http://dx.doi.org/10.1155/2015/124524

Kobo, K. (2021) Africa's plastics bans are pitting the environment against the economy. *Quartz Africa*. Available at: https://qz.com/africa/2007658/why-western-style-plastic-bans-arent-working-in-africa/

Krymowski, J. (2021). California startup turns methane into biodegradable plastic. *The Daily Churn*. Available at: www.darigold.com/mango-materials-biodegradable-methane-plastic/

Lundquist, N.A., Tikoalu, A.D., Worthington, M., Shapter, R., Tonkin, S.J., Stojcevski, F., Mann, M., Gibson, C.T., Gascooke, J.R., Karton, A., Henderson, L.C., Esdaile, L.C., Chalker, J.M. (2020) Reactive compression molding post-inverse vulcanization: A method to assemble, recycle, and repurpose sulfur polymers and composites. *Chemistry – A European Journal*, 10035–10044. https://doi.org/10.1002/chem.202001841

Magoum, I. (2021) AFRICA: 3D printing to turn plastic waste into agricultural tools. Afrik21. Available at: www.afrik21.africa/en/africa-3d-printing-to-turn-plastic-waste-into-agricultural-tools/

Magoum, I. (2022) Kenya: DOW invests in Mr. Green Africa to recycle plastic waste. Afrik21. Available at: www.afrik21.africa/en/kenya-dow-invests-in-mr-green-africa-to-recycle-plastic-waste/

Marinelli, S., Butturi, M.A., Rimini, B., Gamberini, R., Sellitto, M.A. (2021) Estimating the circularity performance of an emerging industrial symbiosis network: The case of recycled plastic fibers in reinforced concrete. *Sustainability*, 13, 10257. https://doi.org/10.3390/su131810257

Medina, M. (2005) Waste picker cooperatives in developing countries. WIEGO/Cornell/SEWA Conference on Membership-Based Organizations of the Poor, Ahmedabad, India. http://dx.doi.org/10.4324/9780203934074.pt4

Metternicht, G., Carr, E., Stafford Smith, M. (2020) Why behavioral change matters to the GEF and what to do about it. A STAP Advisory Document. Scientific and Technical Advisory Panel to the Global Environment Facility. Washington, DC.

Mikula, K., Skrzypczak, D., Izydorczyk, G., Warchol, J., Moustakas, K., Chojnacka, K., Witek-Krowiak. (2021) 3D printing filament as a second life of waste plastics – A review. *Environmental Science and Pollution Research*, 28, 12321–12333. https://doi.org/10.1007/s11356-020-10657-8

Moss, E., Moss & Mollusk Consulting. (2021) Reducing plastic pollution: campaigns that work. Insights and examples to maximize the effectiveness of campaigns for sustainable plastic consumption. *SEI and One Planet*. Available at: www.sei.org/wp-content/uploads/2021/02/210216-caldwell-sle-plastics-report-with-annex-210211.pdf

Ngcuka, O. (2021) South Africa to import plastic waste 'to meet the needs of the industry'. *Daily Maverick*. Available at: www.dailymaverick.co.za/article/2021-09-16-south-africa-to-import-plastic-waste-to-meet-the-needs-of-the-industry/

Nielsen, T.D., Hasselbalch, J., Holmberg, K., Stripple, J. (2019) Politics and the plastic crisis: A review throughout the plastic life cycle. *WIREs Energy Environ*, e360. https://doi.org/10.1002/wene.360

Nyathi, B., Togo, C.A. (2020) Overview of legal and policy framework approaches for plastic bag waste management in African countries. *Journal of Environmental and Public Health*, 2020. https://doi.org/10.1155/2020/8892773

O'Brien, J., Thondhlana, G. (2019) Plastic bag use in South Africa: Perceptions, practices and potential intervention strategies. *Waste Management*, 84, 320–328. https://doi.org/10.1016/j.wasman.2018.11.051

Oghazi, P., Mostaghel, R. (2018) Circular business model challenges and lessons learned – an industrial perspective. *Sustainability*, 10, 739. http://dx.doi.org/10.3390/su10030739

Oteng-Ababio, M (2012) The role of the informal sector in solid waste management in the Gama, Ghana: Challenges and opportunities. *Tijdschrift voor Economische en Sociale Geografie*, 103, 412–425. https://doi.org/10.1111/j.1467-9663.2011.00690.x

Oyinlola, M., Schröder, P., Whitehead, T., Kolade, O., Wakunuma, K., Sharifi, S., Rawn, B., Odumuyiwa, V., Lendelvo, S., Brighty, G., Tijani, B., Jaiyeola, B., Lindunda, L., Mtonga, R., Abolfathi, S. (2022) Digital innovations for transitioning to circular plastic value chains in Africa. *Africa Journal of Management*, 8:1, 83–108. https://doi.org/10.1080/23322373.2021.1999750

Pacini, H., Shi, G., Sanches-Pereira, A., Filho, C. (2021) Network analysis of international trade in plastic scrap. *Sustainable Production and Consumption*, 27, 203–216. https://doi.org/10.1016/j.spc.2020.10.027

Panda, A.K., Singh, R.K., Mishra, D.K. (2010) Thermolysis of waste plastics to liquid fuel: A suitable method for plastic waste management and manufacture of value added products-A world prospective. *Renewable and Sustainable Energy Reviews*, 14, https://doi.org/10.1016/j.rser.2009.07.005

Pant, A., Mostafa, M., Bridle, R. (2020) Understanding the role of subsidies in South Africa's coal-based liquid fuel sector. International Institute for Sustainable Development Policy Brief. Available at: www.iisd.org/system/files/2020-10/subsidies-south-africa-coal-liquid-fuel.pdf

Papaspyrides, C.D., Kiliaris, P. (Eds). (2014) *Polymer Green Flame Retardants*. Elsevier Science Limited, Amsterdam, Netherlands.

Peach, S. (2014) A breakthrough in making plastic from methane? Yale Climate Connections. Available at: https://yaleclimateconnections.org/2014/11/a-breakthrough-in-making-plastic-from-methane/

Persson, L., Almroth, B.M.C., Collins, C.D., Cornell, S., de Wit, C.A., Diamond, M.L., Fantke, P., Hassellöv, M., MacLeod, M., Ryberg, M.W., Jørgensen, P.S., Villarrubia-Gómez, P., Wang, Z., Hauschild, M.Z. (2022) Outside the safe operating space of the planetary boundary for novel entities. *Environmental Science & Technology*, 56, 3, 1510–1521. https://doi.org/10.1021/acs.est.1c04158

Pew Trust. (2020) Breaking the plastic wave. A comprehensive assessment of pathways towards stopping ocean plastic pollution. The Pew Charitable Trust. Available at: www.pewtrusts.org/-/media/assets/2020/07/breakingtheplasticwave_report.pdf

Plastics SA. (2019a) The state of South African recycling companies. Available at: www.plasticsinfo.co.za/2019/08/23/the-state-of-south-african-recycling-companies/

Plastics SA. (2019b) National plastics recycling survey 2018: Recycling is the realization of the plastics circular economy. Available at: www.plasticsinfo.co.za/wp-content/uploads/2019/12/Plastics-Recycling-in-SA-July-2018-Executive-Summary-final.pdf

Preston, F. (2012) A global redesign? Shaping the circular economy. Chatham House Research Paper. Available at: www.chathamhouse.org/2012/03/global-redesign-shaping-circular-economy

Preston, F., Wellesley, L., Lehne, J. (2019) An inclusive circular economy. Priorities for developing countries. Chatham House Research Paper. Available at: www.chathamhouse.org/2019/05/inclusive-circular-economy

Rahimi, A., García, J.M. (2017) Chemical recycling of waste plastics for new materials production. *Nature Reviews Chemistry*, 0046. https://doi.org/10.1038/s41570-017-0046

Rapada, Maria Z., Yu, Derrick E., Yu, Krista D. (2021) Do social media posts influence consumption behavior towards plastic pollution? *Sustainability*, 13, 22, 12334. https://doi.org/10.3390/su132212334

Reddy, M.M., Vivekanandhan, S., Misra, M., Bhatia, S.K., Mohanty, A.K. (2013) Biobased plastics and bionanocomposites: Current status and future opportunities. *Progress in Polymer Science*, 38, 1653–1689. https://doi.org/10.1016/j.progpolymsci.2013.05.006

REMA. (2020) Talking Points for Director General Juliet Kabera High-Level Session of the Geneva Beat Plastic Pollution Dialogues. Rwanda Environment Management Authority. Available at: www.genevaenvironmentnetwork.org/wp-content/uploads/2021/02/Talking-Points-DG-Kabera-High-Level-Session-Geneva-Beat-Plastic-Pollution-Dialogues.pdf

Sadeghi, B., Marfavi, Y., AliAkbari, R., Kowsari, E., Ajdari, F.B., Ramakrishna, S. (2021) Recent studies on recycled PET fibers: Production and applications: A review. *Materials Circular Economy*, 3, 4. https://doi.org/10.1007/s42824-020-00014-y

Schroder, P., Raes, J. (2021) Financing an inclusive circular economy. De-risking investments for circular business models and the SDGs. Chatham House Research Paper. Available at: www.chathamhouse.org/sites/default/files/2021-07/2021-07-16-inclusive-circular-economy-schroder-raes_0.pdf

Steffen, W., Richardson, K., Rockström, J., Cornell, S.E., Fetzer, I., Bennett, E.M., Biggs, R., Carpenter, S.R., de Vries, W., de Wit, C.A., Folke, C., Gerten, D., Heinke, J., Mace, G.M., Persson, L.M., Ramanathan, V., Reyers, B., Sörlin, S., Sustainability. (2015) Planetary boundaries: Guiding human development on a changing planet. *Science*, 347, 1259855. https://doi.org/10.1126/science.1259855

Sun, L., Dong, L., Fang, K., Ren, J., Geng, Y, Fujii, M., Zhang, W., Zhange, N., Liu, Z. (2016) Eco-benefits assessment on urban industrial symbiosis based on material flows analysis and energy evaluation approach: A case of Liuzhou city, China. *Resources, Conservation and Recycling*, 119, https://doi.org/10.1016/j.resconrec.2016.06.007

Taylor, P., Steenmans, K., Steenmans, I. (2020) Blockchain technology for sustainable waste management. *Frontiers in Political Science*, 2, 590923. https://doi.org/10.3389/fpos.2020.590923

Thiounn, T., Smith, R.C. (2020). Advances and approaches for chemical recycling of plastic waste. *Journal of Polymer Science*, 58, 1347–1364. https://doi.org/10.1002/pol.20190261

Tshifularo, C.A., Patnaik, A. (2020) Recycling of plastics into textile raw materials and products. In: Nayak, R. (Ed.) *Woodhead Publishing Series in Textiles, Sustainable Technologies for Fashion and Textiles.* Woodhead Publishing, 311–326. https://doi.org/10.1016/B978-0-08-102867-4.00013-X

Ugorji, C., van der Ven, C. (2021) These 4 methods can help solve Ghana's plastic dilemma. *World Economic Forum.* Available at: www.weforum.org/agenda/2021/09/4-ways-trade-ghana-transition-circular-plastics-economy/

UNEP. (2018) Legal limits on single-use plastics and microplastics: A global review of national laws and regulations. United Nations Environment Program, Nairobi, Kenya.

UNDP. (2019) Plastics and circular economy: Community solutions. United Nations Development Programme, New York.

UNDP Ghana. (2020) Ghana's multi-stakeholder waste recovery initiative. UNDP Ghana. Available at: www.gh.undp.org/content/ghana/en/home/projects/waste_initiative.html

University of Sydney. (2021) Recycling robot could help solve soft plastic waste crisis. Available at: www.sydney.edu.au/news-opinion/news/2021/06/23/recycling-robot-could-help-solve-soft-plastic-waste-crisis-.html

Van Os, E., De Kock, L. (2021) Plastics: From recycling to (post-consumer) recyclate: Industry views on barriers and opportunities in South Africa. WWF South Africa, Cape Town, South Africa.

Veolia. (2019) AfricWaste: The rise of the circular plastics economy in West Africa. Veolia. Available at: www.planet.veolia.com/en/circular-economy-plastics-recycling-abidjan-cote-ivoire-africa

Verghese, K., Jollands, M., Allan, M. (2008) The litterability of plastic bags: Key design criteria. A Report Presented on 5th Australian Conference on Life Cycle Assessment: Achieving Business Benefits from Managing Life Cycle Impacts, Melbourne, 1–10.

WBCSD. (2016) Informal approaches towards a circular economy – Learning from the plastics recycling sector in India. World Business Council for Sustainable Development. Available at: https://sustainable-recycling.org/wp-content/uploads/2017/01/WBCS D_2016_-InformalApproaches.pdf

WEF. (2016) The new plastics economy. Rethinking the future of plastics. World Economic Forum. Geneva. Available at: www3.weforum.org/docs/WEF_The_New_ Plastics_Economy.pdf

WWF. (2018) WWF Plastic File #4: make the circle bigger. Available at: www.wwf.org. za/?26201/plastic-file-04

WEF. (2021) Nigeria joins forces with World Economic Forum to fight plastic pollution. World Economic Forum News Release. Available at: www.weforum.org/press/2021/ 01/nigeria-joins-forces-with-world-economic-forum-to-fight-plastic-pollution/

WWF, Dalberg. (2021) Plastics: The costs to society, the environment and the economy. World Wide Fund for Nature and Dalberg Advisors. Available at: https://media.wwf. no/assets/attachments/Plastics-the-cost-to-society-the-environment-and-the-econ omy-WWF-report.pdf

YeniSafak. (2021) Public-private partnership to tackle harmful plastics in Rwanda. Available at: www.yenisafak.com/en/world/public-private-partnership-to-tackle-harmful-plastics-in-rwanda-3573910

Yeoh, T.N. (2020) A circular economy for all: The case for integrating the informal waste sector in developing countries. Kennedy School Review. Available at: https://ksr.hksp ublications.org/2020/02/17/a-circular-economy-for-all/

3

DIGITAL TECHNOLOGIES AND THE REGIME COMPLEX FOR PLASTICS IN NIGERIA

Natalie Beinisch

1   What Is Regime Complex Theory and Why Does It Matter?

Regime complexes are used to describe the phenomenon of overlapping, non-hierarchical institutions that regulate issue areas at the international level. Its analytical approach is concerned with defining and mapping the boundaries of regulatory spaces and identifying interactions between different regulatory institutions (Raustiala and Victor 2004). The logic of using this approach is that as international rule-making and coordination becomes more complex, power dynamics shift and so too do explanations of why and how rules are made and implemented.

Keohane and Victor (2011) outline three assumptions that shape this approach:

1. International regulatory issues cut across multiple regulatory domains. This is important because there are elements of rule-making that are path dependent and cannot be explained by power dynamics alone.
2. International regulatory problems are complex and usually represent a set of interdependent problems. "Climate Change", for example, includes a number of distinct problems including energy efficiency, transitions to new energies, changes in consumption patterns, etc. Accordingly, there is a greater diversity of interest groups and organisations that participate in regulatory processes, and these organisations may cooperate or compete with one another. This means that examining regulation as interactions between overlapping institutions can be more instructive than studying a single regime.
3. The complexity of rule-making makes forum-shifting possible, but this does not necessarily produce suboptimal regulatory institutions because forum-shifting

DOI: 10.4324/9781003278443-4

allows greater flexibility to manage the complexity and uncertainty associated with international regulatory problems. Defining the overlapping institutions that constitute a regime complex and assessing them against a set of normative criteria (outlined in Section 3) can help to determine whether a bundle of institutions working in a specific issue area are operating in a constructive way or not and can help to identify ways of addressing coordination challenges.

Regime complexes matter because they enable us to study regulation in terms of coordination between organisations and institutions that work in a specific issue area. As the past three decades have ushered in unparalleled growth of regulatory institutions at the international level, the complexity and interdependencies of international regimes are expected to grow.

Interestingly, the regime complex approach is focused on the interdependencies of interstate regulation. Another trend that has taken place alongside the growth of international regimes is growing participation of businesses and civil society organisations in what is referred to as "transnational" regulation (Abbott and Snidal 2009). Abbott (2012) observes that the participation of non-state actors in the development and implementation of regulation has equally produced complex and overlapping institutions that interact with interstate regimes in what he refers to as "conscious parallelism" (Abbott 2012, p. 583). While he underlines that transnational regulation produces even greater diversity and fragmentation of interests and capacity to regulate, Abbott (2012) makes the case for expanding regime complex theory to include analysis of transnational regimes, arguing that both interstate regime complexes and systems of transnational governance "lack clear institutional architectures, yet in both cases organizations and standards are loosely coupled through a common focus" (Abbott 2012, p. 583). Similar observations have been made by Green and Auld (2017) who argue that it is necessary to include private actors in regime complex analysis.

Another point of institutional interaction is between interstate regimes, transnational regimes and national regulation. As nation-states are direct participants in interstate rule-making, they are responsible for implementing or enforcing international regimes at the national level. Nation-states may also be key constituents in transnational regulation and can both shape or be shaped by transnational regulation (Grabosky 2013; Reed et al. 2013; Beinisch 2017; Breslin and Nesadurai 2018; Clapham 2022).

In short, the regime complex approach helps to map and explain growing rule complexity at the international level. While the approach originates from the study of interstate regimes, its core methodological elements, of defining regulatory issue areas, the institutions which participate in standard setting and implementation and the interactions between them, are flexible enough to accommodate a plurality of regulatory forms, including interactions between interstate, transnational and national and sub-national regulation.

2 "Good" and "Bad": Assessing the Function of Regime Complexes

As Tolstoy is famously quoted in the opening page of *Anna Karenina* that "Happy families are all alike; every unhappy family is unhappy in its own way", so too are regime complexes dysfunctional in their own way. Keohane and Victor (2011) outline a set of six normative criteria to assess whether a regime complex is functional or not. These criteria are as follows:

Coherence: The different components of a regime complex may be more or less compatible and mutually reinforcing. The more compatible the components of the regime complex, the more coherent it is.

Accountability: The components of a regime complex should be accountable to their constituents. Constituents include other states, non-government organisations and mass publics. Constituents of the various elements of a regime complex should be well defined and have the right and means to hold others to a set of standards and to impose sanctions if standards have not been met.

Effectiveness: Effectiveness refers to rule appropriateness and compliance. An effective regime is expected to create more net benefits for its constituents.

Determinacy: This refers to the certainty of the meaning of rules. In a highly determinant regime complex, rules and objectives are clear and uncontested as are the pathways to meeting them.

Sustainability: This is equivalent to regime stability and the likelihood that shocks or pressure will disrupt different elements of a regime complex. A more sustainable, or stable, regime is preferred to one that is less stable because it improves certainty about future rules.

Epistemic quality: This refers to consistency between rules and scientific knowledge, capacity to revise rules and accountability of managers who operate regime components.

Keohane and Victor (2011) observe that different elements of regime complexes may vary from highly functional to highly dysfunctional. The degree to which the sum of these elements is functional helps to set our expectations about the overall capacity of a regime complex; however, there are no hard and fast criteria to help distinguish between different degrees of functionality. This is not necessarily problematic as a fixed form of assessment would be less capable of accommodating the diversity of regulatory issues and organisations that make up a regime complex. Indeed, a range of authors have used this approach to evaluate regime functionality including Abbott (2014), Brosig (2013), Nye (2014) and Widerberg and Pattberg (2017). The main purpose of applying these normative criteria to a regime complex is to identify vulnerabilities within a set of loosely coupled regimes and approaches to address them.

3 Regime Complex Theory and Digital Technology

Digitisation is playing an increasingly critical role in policy and regulatory processes. E-government, the delivery of information and services by governments to the public is a near-global phenomenon, with most governments across the world offering some types of digital services to citizens to improve efficiency and facilitate greater regulatory compliance (Fang 2002; West 2007; Jayashree and Marthandan 2010). Within interstate and transnational regulatory systems, digital technologies are likewise deployed in the same manner. For example, online tools such as the NDC Partnership and the NDC reporting tool of the Organisation of African, Pacific and Caribbean States have been created to facilitate and standardise reporting for the Nationally Determined Contributions (NDCs) to meet the goals of the 2015 Paris Agreement on climate change. Transnational regulatory programmes such as the Global Reporting Initiative and the Principles for Responsible Investment are also primarily digitally based.

E-governance, on the other hand, is related to structural changes of governing that are enabled by digital technology (Bannister and Connolly 2012). While Bannister and Connolly (2012) point to examples such as fixmystreet.com where decision-making about road improvements is driven from the bottom-up, they argue that technology has to date been limited in terms of producing meaningful structural transformations. In the same vein, regime complex theory helps to map radical structural changes that impact how transnational issues are governed; however, these changes are linked to changes in views about the role that governments should play in society and the economy (Hood and Dixon 2015; Hood and Rothstein 2000), the growth of the free-trade agenda (Cutler et al. 1999; Cashore et al. 2008) and growing rule density and issue complexity (Alter and Raustiala 2018).

Regime complex theory tends to treat digital technology as a subject rather than a mechanism of regulation or driver of regulatory change, covering issues such as cyber regulation (Raymond 2016; Pawlak 2019), digital trade (Azmeh et al. 2020; Weina 2021), intellectual property rights (Kuyper 2015) and digital sequence information of plant genetic resources (Smyth et al. 2020). However, given that digital technologies have the potential to address issues of accountability, transparency and coordination that are more common in non-hierarchical or "networked" forms of regulation (Newman 2004), the regime complex approach can help us explore the extent to which digital technologies enable new modes of governance.

4 Methodology

This chapter defines and maps the regime complex for plastics in Nigeria and assesses how it functions based on Keohane and Victor's 2011) normative criteria. Based on this assessment, this chapter identifies ways that digital technology may be used to address weaknesses in the regime complex for plastics in Nigeria from an e-government and e-governance perspective.

One of the attractions of regime complex theory is its flexibility. As Gómez-Mera et al. (2020) argue in their review, the "label 'regime complex' is appropriate at any level of analysis as long as institutions under study are analyzed as a set rather than as unconnected units or a cohesive block" (Gómez-Mera et al. 2020, p. 3). It is in this spirit that this chapter builds on the work of Abbott (2012) and incorporates non-state regulatory programmes in its analysis. The next section defines and maps the regime complex for plastics at the international level as well as its relevance at the national and sub-national levels in Nigeria. The mapping is carried out through documentary analysis, based on material that is publicly available. Information about regulatory developments in plastic waste was accessed through literature reviews. The websites and policy documents of regimes identified through the literature reviews were studied and triangulated with news and "grey" research material to find evidence of regime implementation. This constructivist approach is compatible with regime complex theory and allows us to explore a much richer tapestry of regulatory activity, as well as the interactions (or absence of interactions) between them.

Regimes that are identified and mapped are subsequently evaluated against the Keohane and Victor's (2011) normative criteria, which is mapped in Table 3.1. Application of these criteria is more art than science, especially because the details of the ways regulatory systems function are not always obvious from the vantage point of desktop research which is the primary form of data collection for this study. Another methodological challenge is the broad definition this study takes in respect to regimes, meaning there is a high volume of plastic initiatives that qualify as regimes, especially in respect to pledges and commitments of individual businesses. While every regime is different, those which are in abundance, such as business and multi-stakeholder regimes, are evaluated as a group as their characteristics are close enough in similarity that this avoids needless analytical repetition.

Given the observational difficulties of assessing the normative criteria, a best effort approximation is made for each one. Regime coherence is taken as the expression of the core objectives of the regime. These objectives are outlined with an indication as to whether they conflict with other elements of the regime complex. Evidence of established reporting frameworks is used to determine the accountability of regime components. Given the diversity of regimes under the microscope, there is no single reporting framework that can be ascribed as preferential to alternatives, so only observations of whether they can be observed and their type are recorded. The most accessible way to measure rule effectiveness is to determine whether compliance mechanisms are present to reinforce standards. As with reporting, a multitude of soft and hard compliance mechanisms exist; however, binding agreements that can be enforced with a "big stick" are far more likely to mobilise behaviour change, especially when changes are costly (Braithwaite 2006). Determinacy is assessed in this chapter in terms of rule specificity. The assumption in this case is that the more specific the rules and more detail about implementation is clear the more determinant a regime is. The approximation that is used to determine regime sustainability in this chapter

TABLE 3.1 The regime complex for plastics and Nigeria

Domain / Institutional Arrangement	Marine Waste	Safe and Efficient Waste Management and Recovery	Emissions Reductions and Bio-Diversity Management
Multilateral Institutions (IMO, UNEP, UNEA, FAO)	Convention on Prevention of Marine Pollution; International Convention on Prevention of Pollution by Ships; Ad-hoc open ended group on Marine Litter and Microplastics; Ministerial Conference on Marine Litter and Plastics Pollution	Basel Convention: Categorizations of Plastic Waste (May 2019)	
Regional Agreements and Frameworks	Abidjan Convention; Nairobi Convention; EU Directives on Microplastics (multiple, product and process based regulations)	EU Waste Import and Export Restrictions and Bans; European Strategy for Plastics in a Circular Economy; Commonwealth Clean Ocean Alliance	
National and Sub-National Regulation		Material and Material Import Bans (Single Use Plastics, Microbeads, Waste Materials); Extended Producer Responsibility Schemes; National Policy on Plastic Management (Nigeria); Sub-National Plastic Policies; Waste-to-Wealth Schemes and Recycling Programmes; Waste Picker MBOs	

(Continued)

TABLE 3.1 (Continued)

Domain / Institutional Arrangement	Marine Waste	Safe and Efficient Waste Management and Recovery	Emissions Reductions and Bio-Diversity Management
Business Self-Regulation		Material bans, recycling and reuse commitments, material transition commitments	
Multi-Stakeholder Regulation	FAO Code of Conduct for Fisheries	EU Pledging Campaign: Call to for businesses to produce and use more recycled plastics	
		Oceans Plastics Charter	
	Honolulu Strategy on Marine Litter		
		Ellen MacArthur Global Plastic Commitment	
		Alliance to End Plastic Waste	
		Plastic Waste Partnership	
		UK Plastics Pact	
Investor Groups		Coordinated Investor Engagement Frameworks	
		Circulate Capital	
Development Institutions		Recycling value chain initiatives	
		Circular Economy initiatives	

is in terms of institutional development. For example, a regime with an established secretariat, active members and a functioning reporting process is considered to be highly sustainable, while ones that are missing one or more of these functions are categorised as "medium" or "low". Finally, epistemic quality in this study is taken to equate to process. If there is an institutionalised form of rule review and development, epistemic quality is considered to be high; however, if rule review appears to be haphazard or non-exist, then it is classified as medium or low, respectively.

Based on the observations from this assessment, the final section concludes with a discussion about the role that digital technology can play from the perspective of improving delivery and coordination as well as possible structural changes that digitisation may facilitate.

5 Nigeria within the Regime Complex for Plastics

Plastics are ubiquitous in our daily life and are used to produce everything from nappies to cars. However, by many accounts, we are in a period of crisis when it comes to plastic production, use and disposal. Plastic waste accounts for approximately 12% of all solid waste produced on the planet (Ghosh 2020). According to the World Wildlife Fund, 141 million metric tonnes of packaging waste alone was produced in 2015, with only 10% of materials being recycled. The remaining 90% of plastic packaging waste that is generated is landfilled, incinerated or leaked into the environment. Left unchecked, volumes of plastic packaging waste are expected to swell by 40% by the year 2030 (WWF 2019). Rapidly growing volumes of plastic waste have crept into all aspects of natural life: plastic micro-particles can be found in water, arctic snow, soil and in our food (Bergman et al. 2019). By the estimates of one study, the weight of plastic build-up in the ocean will outstrip the collective weight of fish by 2040 (Lau et al. 2020). The impacts of these increasing volumes of plastic waste are not merely academic; they have far-reaching effects, especially for the health of all living populations on earth. Large numbers of marine and animal life have perished from ingesting or getting entangled with plastic materials (Sigler 2014; Gall and Thomson 2015). There have even been reported fatalities of large mammals such as elephants from ingesting plastics (The Guardian 2022).

Even more alarming is the threat posed by waste plastic to natural ecosystems. Research on marine life has found that micro-plastics in water sources affect the endocrine systems of fish, with relatively little known to date about the impacts on humans who consume them (Rochman et al. 2014; Rao 2019; Zhu et al. 2019). The absorption of micro-plastics by animals and plants at the bottom of the food change is also thought to threaten their development and growth (Environmental Investigation Agency 2022).

Another threat of plastic waste comes from open incineration, which is common in countries such as Nigeria that do not have strong waste management infrastructure (Saush and Schulte 2021). Open incineration is a significant health hazard for all forms of life, and the release of toxic chemicals into the atmosphere increases a range of health risks to humans including heart disease, emphysema and damage to the central nervous system (Kawamura and Pavuluri 2010; Verma et al. 2016).

As climate change has become an increasingly salient global agenda item, it is not only issues related to plastic disposal that have sharpened into focus but also there is increasing emphasis on the role that plastic plays in generating greenhouse gas (GHG) emissions and impacting biodiversity and the environment across its life cycle. Importantly, production of virgin plastics depends upon extraction of fossil fuels, and it is estimated that between 4% and 8% of global oil consumption is associated with plastics, with the proportion set to rise to 20% by 2050 (Ellen MacArthur Foundation 2017). Life cycle studies of plastic have also identified that refining and manufacturing plastic materials as well as the proliferation of micro-plastics in water sources play a significant role in the production of GHG emissions, with GHG emissions from plastics expected to reach about 1.34 billion tonnes per year by 2030. This is roughly the equivalent to the emissions produced by 300 new 500-MW coal-fired power plants (Centre for International Environmental Law 2019). Put another way, if the global plastic life cycle were a country, it would be the fifth largest emitter of GHGs in the world (Environmental Investigation Agency 2022).

Concerns about the disposal of plastic, its impact on air quality, animal and human health and marine life as well as growing awareness about the plastic life cycle and its impact on biodiversity and GHG emissions have led to three types of policy action at the international level. The first is related to controlling and managing plastic waste that is leaked into the marine environment, the second is related to improving waste recovery and management systems and the third is focused on reducing production and consumption of plastics.

International efforts to control waste materials including plastics began in the 1970s and instruments such as The Convention on the Prevention of Marine Pollution by Dumping of Wastes and Other Matter and the International Convention for the Prevention of Pollution from Ships were connected primarily to controlling maritime activities which fell under the domain of the International Maritime Organisation. These efforts were followed by regional frameworks such as the Abidjan and Nairobi Conventions, which are facilitated by the United Nations Environmental Programme. The Abidjan and Nairobi Conventions were established in the 1980s and are intended to mobilise and harmonise legal frameworks in Western and Eastern Africa that protect and preserve the marine environment. These frameworks are broader than the conventions which proceeded it because they cover land-based activities that impact the marine environment. Regardless, the conventions serve to coordinate as opposed to enforce or implement regulatory standard setting.

At the national and sub-national levels, plastics have historically fallen under solid waste management regimes which address how waste is segregated and disposed. However, pressure groups in the United Kingdom and Europe have successfully lobbied for specific policies to address plastic waste, including bans on single-use plastics, micro-beads and imports of plastic material. Single-use plastic bans in particular have been successfully diffused across the world since they were introduced in the early 2000s (United Nations Environment Programme 2018). Nigeria is a case in point, having adopted a National Policy on Plastic Management in 2020. The policy is wide-ranging, including a colour-coding scheme for waste separation and consideration of the role of the judiciary in plastic management. However, it also stipulates for a single-use plastic ban to be implemented at the sub-national level. As a bellwether, in Lagos the ban has been applied to staff working at the Lagos State Environmental Protection Agency (Akoni 2022).

Extended Producer Responsibility (EPR) schemes are another important regulatory tool deployed by national governments and have been rolled out since the 1990s, beginning in Northern Europe (Walls 2006). While the design and operations of EPR schemes vary, the key tenet of this approach is that producers bear significant financial or physical responsibility to treat or dispose of waste at its end of life. The view is that this creates incentives for producers to innovate to reduce waste and experiment with new material types in production. EPR schemes cover a range of material types including plastics and have been implemented across the world. A core institutional feature of EPR schemes is Producer Responsibility Organizations (PRO). In Nigeria, the EPR for plastic packaging was initiated by the Nigeria Environmental Standards and Regulation Agency (NESREA) and is implemented by the Food and Beverage Recycling Alliance (FBRA), which was set up in 2013 by the major global Fast Moving Consumer Goods (FMCG) brands operating in Nigeria.

While Nigeria has been quick to adopt progressive policies to restrict plastic use, its institutional context is a far cry from that where these policies originated, which has consequences for implementation. In its most basic terms, public infrastructure to support waste segregation, collection and disposal is weak, with issues such as taxation, transport infrastructure, space and availability of collection and processing agents and facilities undermining efforts to build waste management capacity. Beyond this, however, policy design can be overly ambitious, vague or both as is the case with the Nigerian National Policy on Plastic Management. This makes it difficult to connect policy with responsibilities and resources for implementation.

Another important distinction in the Nigerian context is the critical role of the informal sector in collecting and recycling plastic and other waste material. Much ink has been spilled about the integration and institutionalisation of rights of the informal sector into waste collection and recycling programmes (Chikarmane 2012; Scheinberg 2012; Katusiimeh et al. 2013); however, the informal sector remains largely self-organised, with limited examples of legal or institutional frameworks that provide informal waste pickers with income or health security. In Nigeria, engagement with the informal sector is programmatic and led by state and non-state organisations including local governments, businesses and development organisations, focusing on issues such as price stability, access to off-takers and opportunities to upcycle waste plastics.

Policies that address plastics in terms of emission reductions and biodiversity protection are relatively new, with their origins connected to promotion of "circular economy" principles that are centred on optimising resource use and eliminating waste in economic value chains. This new generation of policy is not separate from marine protection or waste management agendas but rather seeks to add new dimensions that are focused on reducing or eliminating the production of plastics all together.

The European Union (EU) has established the most comprehensive and stringent policy framework to date with its Strategy for Plastics in a Circular Economy. The strategy has facilitated new sets of directives that set, among other things, standards for recycled components in materials. As a part of the strategy, the EU has set targets for plastic recycling and established the EU Pledging Campaign, a voluntary programme for businesses to produce and use more recycled plastics and initiated amendments to the Basel Convention, which covers transboundary movement and disposal of hazardous wastes, to include plastic waste in a bid to prevent dumping.

Much of the new generations of policy initiatives that incorporate circular economy principles are developed within what is described as "the Governance Triangle" (Abbott and Snidal 2009) of States, Businesses and Civil Society Organisations. This includes the Global Plastic Commitment launched by the Ellen MacArthur Foundation, a UK-based advocacy organisation, and the Alliance to End Plastic Waste, set up by polymer producing organisations. As with the example of the EU Pledging Campaign, action mobilising non-state actors is driven by governments as well.

Table 3.1 maps the existing regime complex for plastics based on the policy domains and the type of institutional arrangement driving policy.

Issues related to plastic production, consumption and disposal cut across issues of marine protection, waste management, health and safety, biodiversity and climate change, with plastic management entering existing frameworks such as the Basel Convention and appearing on the agenda of the United Nations Environmental Assembly through conferences and working groups on marine litter. As circular economy approaches have popularised, production, consumption and disposal of plastics have increasingly been addressed in conjunction with one another, creating overlaps between previously elemental regimes that were focused either on marine protection or waste management. There is also a proliferation of multi-stakeholder agreements, some of which commit businesses to eliminate or reduce plastic waste in their value chains, and these are complemented by individual commitments by businesses. Increasing issue and institutional overlap as well as rule complexity means that international frameworks governing plastics bear the hallmarks of a regime complex.

The regime complex for plastics is observed in multiple forms in Nigeria. The first is through regulatory diffusion such as plastic bans and EPR schemes. The second is through business implementation of both statutory and voluntary plastic commitments, which is most visible through programmatic recycling and collection schemes and growing demand for recycled polyethylene terephthalate (rPET) exports. The third is via development institutions seeking to build institutional capacity to address waste reduction, waste management and recycling. Consistent with the picture painted at the international level, the regime complex for plastics in Nigeria involves multiple institutional arrangements; however, the primary focus of activity is on waste management and recovery.

6 Assessing the Regime Complex for Plastics

Table 3.2 provides an overview of the normative assessment for the regime complex for plastics. At the international level, regimes are broadly coherent. There is no real evidence that membership or compliance to one regime conflicts or contradicts with compliance to another. This limits risks of forum-shifting as new forms of regimes that materialise appear to operate in the spirit of "conscious parallelism" described by Abbott (2012). However, at the national level, there does appear to be inconsistencies with international policies that have been adopted by the federal government and existing legislation, mainly because technical specifications are not updated to facilitate recycling at the pace in which new policies are introduced.

In terms of accountability, reporting is varied with more established international and regional frameworks having very institutionalised reporting processes and others being more dynamic. This variety is a function of the maturity and orientation of some regimes studied. For example, the Ad-hoc Open-Ended Group on Marine Litter and Micro-plastics exists more to establish international practices than to enforce them. Partnerships, alliances and business regulation present a similar case where the maturity and orientation of rules is varied. In this respect, a standardised reporting framework is not necessarily expected or desired. In other cases, such as the Abidjan and Nairobi Conventions, reporting frameworks exist but are not used systematically. In Nigeria, reporting is patchy due to slow implementation of regimes.

One of the loudest calls to action in the past year has been for a binding treaty on plastic (Centre for International Environmental Law 2021). As EU Directives on plastics are the only binding regime that constrain plastic production, use and consumption at a multi-country level, this sense of urgency is logical. Unfortunately, the evidence from this study indicates that regardless of whether a treaty is binding or not, there are significant headwinds in terms of how countries like Nigeria are positioned to implement binding international rules. This may be due in part to the way that international regimes have historically been translated into policy in Nigeria, as from the perspective of rule determinacy, rules on EPR and plastic management have been relatively vague and far-reaching, making them very difficult to translate into practicable law. From the perspective of sustainability and epistemic quality, the challenges are comparable as there is little evidence of institutional capacity or process to implement regulation

TABLE 3.2 Normative assessment of the regime complex for plastics

Regime ——————— Normative Criteria	Basel Convention: Categorizations of Plastic Waste (May 2019)	Ad-hoc open ended group on Marine Litter and Microplastics	Abidjan and Nairobi Conventions	EU Plastic Directives
Coherence *Different components of a regime complex may be more or less compatible and mutually reinforcing. The more compatible the components of regime complex, the more coherent it is.*	Evidence of a "ratcheting up" of standards, consistent with goals to improve waste collection and recycling rates globally and reduce dumping of plastic materials.	Consistent with "Circular Economy" approaches, emphasizes integration of life-cycle approaches to existing marine protection and waste management frameworks. Emphasis on technical expertise and multi-stakeholder approaches that are endorsed by states.	Emphasis on marine and coastal protection, regional coordination.	Eliminates and restricts use of plastics materials considered most threatening to the environment. Reinforces EPR schemes.
Accountability *Constituents of the various elements of a regime complex should be well defined and have the right and means to hold others to a set of standards and to impose sanctions if standards have not been met.*	Standardized National Reports for member states	Meeting-based reporting	Programmatic, limited reporting mechanisms	Reporting and labelling. Directives implemented by members states.
Effectiveness *Rule appropriateness and compliance. An effective regime is expected to create more net benefits for its constituents.*	Non-Binding	Non-Binding	Non-Binding	Binding
Determinancy *Certainty of the meaning of rules. In a highly determinant regime complex, rules and objectives are clear and uncontested as are the pathways to meeting them.*	Detailed and Specific	Exploratory	Thematic	Detailed and Specific

Plastic Waste Partnerships and Alliances	Business Material Bans	National Material and Material Import Bans	Nigeria Plastics Management Policy	Sub-National Waste Management Frameworks	Extended Producer Responsibility (Nigeria)
Consistent with national and regional measures to improve recycling and innovation to reduce waste. Ellen MacArthur Plastic Commitment focuses	Bans/Material phase-outs vary depending upon company. Generally consistent with regional and multistakeholder frameworks.	International standards such as single use plastic bans, amendments to the Basel Convention are being integrated or introduced into national policy frameworks.	Objective is consistent with "Circular Economy" approaches, content of policy is broad, ambitious.	Approach to plastics is programmatic, consistent with national and international objectives, however there can be conflicts with historical waste management regulations and plastic reduction/recycling ambitions.	EPR guidelines were introduced by the Nigerian Environmental Standards Agency in 2014. The guidelines require companies to submit individual EPR plans to the agency. The principles of the EPR are consistent with EPR schemes globally.
Varying reporting frameworks.	Non-binding, varying reporting frameworks.	Not yet implemented in Nigeria	Not fully implemented	N/A	Reporting to NESREA through individual plans.
Non-binding.	Non-binding	Not yet implemented in Nigeria	Scope of policy too broad to be enforced.	N/A	Technically binding.
Varies	Varies	Not yet implemented in Nigeria	Broad and Vague	N/A	General framework and guidance.

(Continued)

TABLE 3.2 (Continued)

Regime ————————— Normative Criteria	Basel Convention: Categorizations of Plastic Waste (May 2019)	Ad-hoc open ended group on Marine Litter and Microplastics	Abidjan and Nairobi Conventions	EU Plastic Directives
Sustainability *Likelihood that shocks or pressure will disrupt different elements of a regime complex. A more sustainable, or stable regime is preferred to one that is less stable because it improves certainty about future rules.*	High	Low	Medium	High
Epistemic Quality *Consistency between rules and scientific knowledge, capacity to revise rules and accountability of managers that operate regime components*	High	High	Medium	High

derived from international regimes. This is not to say that the effect of a binding international treaty would be negligible in Nigeria. Rather, based on this analysis, the expectation is that its translation would be mediated by domestic regulatory institutions that lack competencies and resources to implement them, so the most likely sources of interaction of Nigerian economic actors with such a treaty would come from international actors that have obligations or interests to implement it.

7 Digital Technology and the Regime Complex for Plastics

The previous section paints a picture of the regime complex for plastics where there is broad consistency about the goals of addressing plastic waste and high levels of variation in terms of how different regimes implement and report on these goals and address compliance. This variation stems from three factors.

The first is maturity: while there are some regimes where rules and processes are well established, in others, details about specific rules and compliance are still being worked out. The second has to do with orientation and the expansive definition this study takes of "regimes". Not all regimes in this study are focused on constraining or controlling behaviour. Some, such as the End Plastic Waste Alliance are focused rather on exploring and investing in alternatives or capabilities that reduce plastic and plastic waste.

Plastic Waste Partnerships and Alliances	Business Material Bans	National Material and Material Import Bans	Nigeria Plastics Management Policy	Sub-National Waste Management Frameworks	Extended Producer Responsibility (Nigeria)
Low	Low	Low	Low	Low	Medium
Varies	Varies	Low	Low	Low-Medium	Low-Medium

The third has to do with institutional development. Not all regimes surveyed demonstrated evidence of rule implementation. This is particularly the case for the Nigerian regimes included in this study. This is not especially insightful or interesting as a finding, as the limits of the regulatory capacity of developing countries have been a topic at least in scholarly and activist literature since the 1990s (Strange 1996; Braithwaite 2006; Bartley 2010; Jia et al. 2018).

However, this observation does help to shape questions about the potential role that digitisation can play in terms of building institutional capabilities to improve the overall function of a regime complex (or perhaps reduce its dysfunction). For better or worse, the single most glaring issue in this study is the consistency between international aspirations on plastics and those articulated in plastic-related policies in Nigeria, which is contrasted by limited domestic regulatory progress. This challenge is observed across every normative criterion more so than any other regime that was included.

Are there ways that digital technology can substitute or complement state regulatory capacity? The answer is not that easy. In areas like standard setting, rules are context dependent, and knowledge of the institutional environment is as important as technical standards. Online databases may help to improve knowledge of detailed technical standards and their implementation, but there are rare cases where regulation can simply be "plug and played". There is evidence of this in the Nigerian case, as international standards have been integrated into federal and state policy and legislative frameworks, but the outputs are very much different than observed in other markets. This phenomenon is not unique to Nigeria, nor to developing economies. Indeed, sociology of law and legal positivist approaches make the relationship between legal structure and legal practice the primary objective of their study, underlining there is no fixed relationship between rules and their implementation.[1]

From an e-government perspective, digital technology could be leveraged to improve the monitoring of plastic exports to Nigeria. This information is already captured through the United Nations Comtrade Database (Babayemi et al. 2018); however, consistent and standardised reporting requirements for exporting countries that are laid out in a statutory international framework could strengthen this further and help to build a better picture of long-term consumption and waste management practices.

Digital technology could also be used to centralise and standardise data collection on plastic waste and recovery. The assessment in this chapter illustrates that there are multiple programmatic initiatives supporting waste management and recycling, but very little is known in aggregate about collection and conversion rates or pricing. One of the most promising technologies to emerge to address this issue is blockchain because it enables transactions and decision-making to take place in a decentralised way. From a governance perspective, there are substantial implications, as blockchain is believed to be capable of replacing or complement contract-based or relational governance systems (Keller et al. 2021; Lumineau et al. 2021), especially where multiple contracts or transactions take place (Dasaklis et al. 2022). While programmes making use of blockchain technology are fairly recent, such a tool could be used to improve coordination among programme funders and greater incentives for participants in the recycling industry to drive reporting standards.

Based on the assessment of this chapter however, the scenario where digital technology replaces institutional capacity so that we enter an era of e-governance appears to be a dream of a more distant future. The most obvious uses of digital technology to improve the normative dimensions of the regime complex for plastics are related to improving information exchange and reliability. These applications would still need to be led and implemented by organisations participating in this complex. The application of digital technology to enhance rule quality and specificity is also more limited as are the uses of technology to enhance rule compliance. Thus, while digital technologies offer tremendous hope in terms of addressing coordination challenges through improved information flows, we are not yet at a stage where e-governance is feasible.

Note

1 See, for example, Braithwaite, John, and Peter Drahos. *Global Business Regulation.* Cambridge University Press, 2000; Piore, Michael J. "Beyond Markets: Sociology, street-level bureaucracy, and the management of the public sector." *Regulation &*

Governance 5.1 (2011): 145–164. Berman, Mitchell N. "How Practices Make Principles, and How Principles Make Rules." *U of Penn Law School, Public Law Research Paper* 22-03 (2022).

References

Abbott, Kenneth W. "The transnational regime complex for climate change." *Environment and Planning C: Government and Policy* 30.4 (2012): 571–590.

Abbott, Kenneth W. "Strengthening the transnational regime complex for climate change." *Transnational Environmental Law* 3.1 (2014): 57–88.

Abbott, Kenneth W., and Duncan Snidal. "Chapter Two: The Governance Triangle: Regulatory Standards Institutions and the Shadow of the State." In Mattli, W. and Woods, N. (eds) *The Politics of Global Regulation*. Princeton University Press, 2009, pp. 44–88.

Akoni, Olasunkami. "LASEPA launches ban on single-use plastics, pet bottles, others among staff." Vanguard News. Available at www.vanguardngr.com/2022/01/lasepa-launches-ban-on-single-use-plastics-pet-bottles-others-among-staff/. January 5, 2022.

Alter, Karen J., and Kal Raustiala. "The rise of international regime complexity." *Annual Review of Law and Social Science* 14 (2018): 329–349.

Azmeh, Shamel, Christopher Foster, and Jaime Echavarri. "The international trade regime and the quest for free digital trade." *International Studies Review* 22.3 (2020): 671–692.

Babayemi, Joshua O., et al. "Initial inventory of plastics imports in Nigeria as a basis for more sustainable management policies." *Journal of Health and Pollution* 8.18 (2018): 180601.

Bannister, Frank, and Regina Connolly. "Defining e-governance." *e-Service Journal: A Journal of Electronic Services in the Public and Private Sectors* 8.2 (2012): 3–25.

Bartley, Tim. "Transnational private regulation in practice: The limits of forest and labor standards certification in Indonesia." *Business and Politics* 12.3 (2010): 1–34.

Beinisch, Natalie. *Making it work: The development and evolution of transnational labour regulation*. Diss. The London School of Economics and Political Science (LSE), 2017.

Bergmann, Melanie, et al. "White and wonderful? Microplastics prevail in snow from the Alps to the Arctic." *Science Advances* 5.8 (2019): eaax1157.

Berman, Mitchell N. "How Practices Make Principles, and How Principles Make Rules." *U of Penn Law School, Public Law Research Paper* 22.03 (2022).

Braithwaite, John. "Responsive regulation and developing economies." *World Development* 34.5 (2006): 884–898.

Breslin, Shaun, and Helen E.S. Nesadurai. "Who governs and how? Non-state actors and transnational governance in Southeast Asia." *Journal of Contemporary Asia* 48.2 (2018): 187–203.

Brosig, Malte. "Introduction: The African security regime complex—exploring converging actors and policies." *African Security* 6.3–4 (2013): 171–190.

Cashore, Benjamin, Graeme Auld, and Deanna Newsom. *Governing Through Markets*. Yale University Press, 2008.

Centre for International Environmental Law. Plastic & climate: The hidden costs of a plastic planet. Available at www.ciel.org/plasticandclimate/. 2019.

Centre for International Environmental Law. Over 700 groups call for an international plastics treaty. Available at www.ciel.org/news/over-700-groups-call-for-an-international-plastics-treaty/ December 14, 2021.

Chikarmane, Poornima. "Integrating waste pickers into municipal solid waste management in Pune, India." *WIEGO Policy Brief (Urban Policies)* 8 (2012): 23.

Clapham, Andrew. *Non-State Actors*. Edward Elgar Publishing Limited, 2022.

Cutler, A. Claire, Virginia Haufler, and Tony Porter, eds. *Private Authority and International Affairs*. Suny Press, 1999.

Dasaklis, Thomas K., et al. "A systematic literature review of blockchain-enabled supply chain traceability implementations." *Sustainability* 14.4 (2022): 2439.

Ellen MacArthur Foundation. The new plastics economy: Rethinking the future of plastics. Available online at: www.ellenmacarthurfoundation.org/publications/the-new-plastics-economy-rethinking-the-future-of-plastics, 2017.

Environmental Investigation Agency. Connecting the dots: Plastic pollution and the planetary emergency. Available at: https://eia-international.org/wp-content/uploads/2022-EIA-Report-Connecting-the-Dots-SPREADS.pdf. 2022.

Fang, Zhiyuan. "E-government in digital era: Concept, practice, and development." *International Journal of the Computer, the Internet and Management* 10.2 (2002): 1–22.

Gall, Sarah C., and Richard C. Thompson. "The impact of debris on marine life." *Marine Pollution Bulletin* 92.1–2 (2015): 170–179.

Ghosh, Sadhan Kumar, ed. *Circular Economy: Global Perspective*. Springer, 2020.

Gómez-Mera, Laura, Jean-Frédéric Morin, and Thijs Van de Graaf. "Regime complexes." In Biermann, Franck and Rakhyun Kim (eds) *Architectures of Earth System Governance: Institutional Complexity and Structural Transformation*. Cambridge University Press, 2020, pp. 137–157.

Grabosky, Peter. "Beyond responsive regulation: The expanding role of non-state actors in the regulatory process." *Regulation & Governance* 7.1 (2013): 114–123.

Green, Jessica F., and Graeme Auld. "Unbundling the regime complex: The effects of private authority." *Transnational Environmental Law* 6.2 (2017): 259–284.

Hood, Christopher C., and Henry Rothstein. "Business risk management in government: Pitfalls and possibilities" (October 2000). CARR Discussion Paper No. 0 (Launch Paper). Available at: SSRN: http://dx.doi.org/10.2139/ssrn.471221

Hood, Christopher, and Ruth Dixon. "What we have to show for 30 years of new public management: Higher costs, more complaints." *Governance* 28.3 (2015): 265–267.

Jayashree, Sreenivasan, and Govindan Marthandan. "Government to E-government to E-society." *Journal of Applied Sciences (Faisalabad)* 10.19 (2010): 2205–2210.

Jia, Fu, et al. "Sustainable supply chain management in developing countries: An analysis of the literature." *Journal of Cleaner Production* 189 (2018): 263–278.

Katusiimeh, Mesharch W., Kees Burger, and Arthur P.J. Mol. "Informal waste collection and its co-existence with the formal waste sector: The case of Kampala, Uganda." *Habitat International* 38 (2013): 1–9.

Kawamura, K., and C. M. Pavuluri. "New directions: Need for better understanding of plastic waste burning as inferred from high abundance of terephthalic acid in South Asian aerosols." *Atmospheric Environment* 44.39 (2010): 5320–5321.

Keller, Arne, et al. "Alliance governance mechanisms in the face of disruption." *Organization Science* 32.6 (2021): 1542–1570.

Keohane, Robert O., and David G. Victor. "The regime complex for climate change." *Perspectives on Politics* 9.1 (2011): 7–23.

Kuyper, Jonathan. "Deliberative capacity in the intellectual property rights regime complex." *Critical Policy Studies* 9.3 (2015): 317–338.

Lau, Winnie W.Y., et al. "Evaluating scenarios toward zero plastic pollution." *Science* 369.6510 (2020): 1455–1461.

Lumineau, Fabrice, Wenqian Wang, and Oliver Schilke. "Blockchain governance—A new way of organizing collaborations?" *Organization Science* 32.2 (2021): 500–521.

Newman, Janet. "Constructing accountability: Network governance and managerial agency." *Public Policy and Administration* 19.4 (2004): 17–33.

Nye, Joseph S. *The Regime Complex for Managing Global Cyber Activities.* Vol. 1. Belfer Center for Science and International Affairs, John F. Kennedy School of Government, Harvard University, 2014.

Pawlak, Patryk. "The EU's role in shaping the cyber regime complex." *European Foreign Affairs Review* 24.2 (2019): 167–186.

Piore, Michael J. "Beyond markets: Sociology, street-level bureaucracy, and the management of the public sector." *Regulation & Governance* 5.1 (2011): 145–164.

Rao, B. Madhusudana. "Microplastics in the aquatic environment: Implications for post-harvest fish quality." *Indian Journal of Fisheries* 66.1 (2019): 142–152.

Raustiala, Kal, and David G. Victor. "The regime complex for plant genetic resources." *International Organization* 58.2 (2004): 277–309.

Raymond, Mark. "Engaging security and intelligence practitioners in the emerging cyber regime complex." *The Cyber Defense Review* 1.2 (2016): 81–94.

Reed, Ananya Mukherjee, Darryl Reed, and Peter Utting, eds. *Business Regulation and Non-state Actors: Whose Standards? Whose Development?* Vol. 93. Routledge, 2013.

Rochman, Chelsea M., et al. "Early warning signs of endocrine disruption in adult fish from the ingestion of polyethylene with and without sorbed chemical pollutants from the marine environment." *Science of the Total Environment* 493 (2014): 656–661.

Saush, Anuj and Uwe Schulte. "Plastics Solid Waste Management." The Conference Board Plastics Working Group. Available at www.conference-board.org/topics/plastic/plastic-solid-waste-management. April 2021.

Scheinberg, Anne. "Informal sector integration and high performance recycling: Evidence from 20 cities." *Women in Informal Employment Globalizing and Organizing (WIEGO), Manchester* 23 (2012).

Sigler, Michelle. "The effects of plastic pollution on aquatic wildlife: Current situations and future solutions." *Water, Air, & Soil Pollution* 225.11 (2014): 1–9.

Smyth, Stuart J., et al. "Implications of biological information digitization: Access and benefit sharing of plant genetic resources." *The Journal of World Intellectual Property* 23.3–4 (2020): 267–287.

Strange, Susan. *The Retreat of the State: The Diffusion of Power in the World Economy.* Cambridge University Press, 1996.

The Guardian. "Two more elephants die after eating plastic waste in Sri Lankan dump." Available at: www.theguardian.com/world/2022/jan/14/two-more-elephants-die-after-eating-plastic-waste-in-sri-lankan-dump. 14 January 2022.

United Nations Environment Programme. "Single use plastics: A roadmap for sustainability." Available at: https://wedocs.unep.org/bitstream/handle/20.500.11822/25496/singleUsePlastic_sustainability.pdf. 2018.

Verma, Rinku, et al. "Toxic pollutants from plastic waste-a review." *Procedia Environmental Sciences* 35 (2016): 701–708.

Walls, Margaret. "EPR Policies and Product Design: Economic theory and selected case studies." *Working group on waste prevention and recycling. Environment Directorate. Environmental Policy Committee* (2006).

Weina, Lai. "E-conomic agreement: coherence and contestation in the emerging regime complex of preferential trade agreements governing digital trade in the Asia Pacific" (2021). Undergrad Thesis Available at: https://scholarbank.nus.edu.sg/handle/10635/199477.

West, Darrell. "Global e-government, 2007." (2007). Providence, RI: Center for Public Policy, Brown University.

Widerberg, Oscar, and Philipp Pattberg. "Accountability challenges in the transnational regime complex for climate change." *Review of Policy Research* 34.1 (2017): 68–87.

WWF. "Solving plastic pollution through accountability." *WWF—World Wide Fund for Nature, Gland, Switzerland* (2019).

Zhu, Lin, et al. "Microplastic ingestion in deep-sea fish from the South China Sea." *Science of the Total Environment* 677 (2019): 493–501.

4

FROM POLYMERS TO MICROPLASTICS

Plastic value chains in Africa

Patrick Schröder and Muyiwa Oyinlola

1 Introduction

According to the World Bank, plastic waste accounted for 12% of all municipal solid waste globally in 2016. East Asia and the Pacific accounted for 57 million tonnes of the total 242 tonnes of plastic waste, Europe and Central Asia accounted for 45 million tonnes, and North America accounted for 35 million tonnes. Only 17 million tonnes of plastic waste were generated in Africa (Kaza et al., 2018). The total worldwide production of plastics in 2020 amounted to some 367 million metric tonnes (Statista, 2022). Even though sub-Saharan Africa currently accounts for the lowest proportion of plastic waste globally (Ayeleru et al., 2020), growing population, changes in consumption and lifestyle trends as well as increased urbanisation are expected to increase the plastic waste generated in Africa. Furthermore, the ability to provide low-cost hygienic packaging implies that its use will increase as various sectors such as food and beverage grow (Narancic and O'Connor, 2019). Consequently, it is anticipated that Africa will be significantly impacted by the global waste crisis. This is because the use of plastics as packaging materials, which represents over a third of plastics produced, will result in consumer behaviours changing to a "throw-away culture". In other words, consumers will move from reusable to single-use containers that are disposed within a short timeframe, thereby increasing the contribution to municipal solid waste (Jambeck et al., 2015). In fact, the continent is anticipated to experience almost 200% increase in waste generated by 2050, with much of this being plastic (Kaza et al., 2018). This is likely to pose significant environmental and health challenges if not managed properly. Lebreton and Andrady (2019) predicted that the amount of mismanaged plastic in Africa will be disproportionately high unless significant investments in waste management infrastructure are made.

DOI: 10.4324/9781003278443-5

Consequently, several studies have highlighted the need to stop leakage of plastic materials into the environment across the entire plastic value chain, not just at the end-of-life stage. Geyer et al. (2017) note that an important consideration in Africa should be the control of the entire value chain as plastics enter the ecosystem from various entry points, and Ayeleru et al. (2020) highlighted the need for more academic studies to focus on mitigating plastic leakage, particularly in sub-Saharan Africa. The value chain is especially crucial for the end-of-life stage because less than 5% of plastic waste in Africa gets recycled (UNEP, 2018), with the remainder disposed of through unregulated landfills, open burning, open dumping and dumping into water bodies. Furthermore, due to a lack of robust waste management infrastructure, the majority of collection is done by the informal sector, which consists of rubbish pickers who are insufficiently resourced to satisfy the demand (Joshi et al., 2019).

This chapter therefore examines the plastic value chain in Africa, specifically exploring the production, import, use and end-of-life stages. It further illustrates how digital tools and technologies could help in minimising leakage as well as improving material flow through the value chain. This is significant because systemic solutions to the plastic pollution challenge are only possible if a life cycle perspective is embraced.

2   Plastic value chains in Africa

Although the current data and information available on the plastic value chain in Africa is limited, this section draws on the extant literature and online data resources to give insights on the production, import, use and disposal of plastics in Africa.

2.1   Production

The production of plastics on the African continent is significantly lower compared to other regions (Babayemi et al., 2019). The combined production of the top eight producing countries was 15 Mt between 2009 and 2015 (Babayemi et al., 2019). Africa together with the Middle East accounted for only 7% of the global plastic material production in 2020, compared to China which accounted for 32% and the North American Free Trade Area (NAFTA), the world's second-largest producer of plastics, accounting for 19% of worldwide output. Oil and liquefied natural gas (LNG) are the primary plastic production feedstocks in Africa; nevertheless, coal is a key feedstock in South Africa (Mofo, 2020). According to Mofo (2020), a transition to more localised value chains and greater use of natural gas feedstock will provide an opportunity to enhance value for national industries while decreasing environmental consequences.

The top three producers of plastic resin in Africa are Egypt, South Africa and Nigeria who were estimated to have, respectively, produced 2329 kt, 1410 kt, 513 kt of plastic in 2020 (Euromap, 2020). In these countries as well as others across the continent, the production of plastic packaging accounts for between 50% and 60% of the total consumption (Euromap, 2020). This is fuelled by large multinational companies who manufacture and sell fast-moving consumer goods across Africa (Break Free From Plastic, 2019). The drinking water supply chain and sachet water packaging are also significant contributors, with sachet water packaging being one of the most significant aspects of plastic waste seen in Africa. This pollution generated by the brands has contributed to other difficulties such as blocked city drainage, mosquito breeding and localised flooding (Williams et al., 2019).

Similarly, the dominant processing method for plastic production is by extrusion, accounting for above 50% of the processing in most countries. Other methods such as injection moulding, expanded polystyrene (EPS) foam moulding, polyethylene terephthalate (PET) preform and stretch blow moulding account for much less (Euromap, 2020). Table 4.1 shows the number of companies that manufacture plastic products. These are involved in the processing of new or spent (i.e., recycled) plastic resins into a wide range of intermediate or finished plastic products employing methods such as compression moulding, extrusion moulding, injection moulding, blow moulding and casting. The majority of these businesses are in Northern Africa, particularly Morocco and Algeria.

A report by the Sustainable Manufacturing and Environmental Pollution Programme (SMEP), published by the Stockholm Environment Institute showed plastics, although not the main polluting sector in African manufacturing, make a significant contribution (SEI and UoY, 2020).

Data on production of plastic from biomass is limited, although Oyinlola et al. (2022b) suggest that there are pockets of innovations across the continent which have used biomass such as banana peels and seaweed to create plastic products. This alternate approach will reduce the need for raw material extraction, as well as save on the millions of tonnes of chemicals during monomer or polymer production.

TABLE 4.1 Plastic manufacturing companies in Africa

Country	No. of companies
Morocco	2091
Algeria	5002
South Africa	422
Tunisia	850
Egypt	68
Nigeria	50
Ghana	26
Tanzania	22
Ivory Coast	23
Uganda	26
Ethiopia	15

(*Source*: D&B Hoovers, 2022)

In recent times, plastic producers have been working to minimise leakage from the value chain, for example, in a bid to address the end of life of the plastic value chain, multinationals such as Nestle, Coca-Cola and Pepsi launched the "African Plastics Recycling Alliance" in 2019 (Break Free From Plastic, 2019). This alliance aims to improve the plastic recycling infrastructure across sub-Saharan Africa (IISD, 2019). Another intervention that producers will use to prevent leakage in the value chain is the extended producer responsibility (EPR) scheme where the producers will be directly responsible for recovery of post-consumer waste.

2.2 Imports

In order to meet the expanding demands of a fast-growing middle-class economy, the continent is also a big importer of plastic polymers and plastic items. According to Babayemi et al. (2019), the 54 African countries imported an estimated 172 Mt of polymers and plastics worth $285 billion between 1990 and 2017. In addition, product components were imported, totalling an estimated 230 Mt of plastics. Egypt (18.4%), Nigeria (16.9%) and South Africa (11.6%) were the top three importing countries.

Figure 4.1 shows imports of plastic commodities to Africa (CircularEconomy. Earth, 2022). It can be observed that the highest trade flows occur between China to Kenya (7.1 kt), Japan to Nigeria (5.3 kt), China to Nigeria (4.4 kt), Thailand to Nigeria (4.3 kt) and China to Angola (2.4 kt). The top importing African countries include Nigeria (17.7 kt), Kenya (10.3 kt) and South Africa (10 kt), while the top exporters to Africa include China (21.9 kt), Japan (7.2 kt) and the United States (5.5 kt).

The import of plastic waste onto the continent is another contributor to the plastic value chain. This was aggravated in 2018 when China prohibited the importation of many kinds of plastic waste. This action prompted countries such

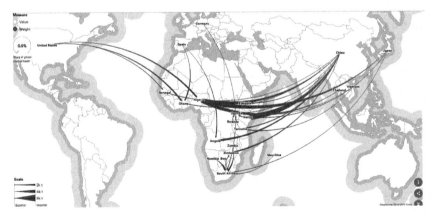

FIGURE 4.1   Plastic waste imports to Africa in the year 2020

(*Source:* CircularEconomy. Earth, 2022)

as the United States, the United Kingdom, Europe and Japan, which rely on shipping plastic waste to China, to consider other places across the world, especially African countries, as an alternate destination. As a result, plastic waste exports to Africa quadrupled in 2019 compared to the previous year (Tabuchi et al., 2020). Ethiopia, Ghana, Kenya, Senegal, South Africa, Tanzania and Uganda were among the nations that received plastic waste from the United States, the majority of which was abandoned or burned (Lerner, 2020).

In their investigation of global plastic waste trade networks, Pacini et al. (2021) discovered that Africa is often underrepresented in international networks, which appears to be attributable to lower plastic usage, informal trade and data reporting difficulties. According to UN Comtrade data, African nations received 82.1 kt of plastic waste and scrap imports in 2019, accounting for around 1% of global plastic waste trade. This represents a very minor portion of total worldwide plastic waste commerce; nonetheless, there was likely more plastic waste traded than was officially recorded.

According to Interpol data, there has been an alarming surge in illegal plastic pollution trade worldwide since 2018. To conceal the origin of the waste shipment, plastic waste is typically redirected to Southeast Asia via various transit countries. There is currently no evidence of criminal trends in plastic waste in Africa, such as unlawful shipment of plastic waste. The analysis, however, suggested a considerable number of unlawful e-waste trade channels, which might potentially be exploited for illegal plastic trading. Some African countries are already receiving substantial amounts of "soon-to-be waste" plastic material embedded in illegally imported e-waste (INTERPOL, 2020).

2.3 Use

The use phase of plastics is responsible for a significant proportion of the leakage from the value chain stemming mainly from households, open markets, formal institutions, public and commercial areas and the manufacturing companies (Kaseva and Mbuligwe, 2005). The United Nations (UN) reports that about 99% are used for less than six months (Ayeleru et al., 2020). Babayemi et al. (2019) suggested that the plastic consumption per capita in Africa stood at about 16 kg per year in 2015 which is low compared to other regions. Plastic consumption in sub-Saharan Africa is expected to be more than six times higher in 2060, according to the Organisation for Economic Co-operation and Development (OECD), due to significant economic and demographic growth (OECD, 2022). The bulk of these usages can be attributed to fast-moving consumer goods and packaging; however, there is a significant use in other sectors such as building and construction, electrical and electronics, textiles, medical services and transport (Narancic and O'Connor, 2019).

The affordability of plastics makes them a suitable candidate for alternate delivery mechanisms in Africa, for example, in catering for the lack of water infrastructure across Africa, plastics have been used to make portable water available to various socioeconomic levels such as low-cost low-density polyethylene (LDPE) sachet packaging for low-income communities as well as PET plastic containers used in corporate environments. Similarly, plastics have been used for hygienic transport of food to prevent food losses and wastage.

The main driver of leakage with the use phase is the lack of adequate collection and disposal infrastructure. For example, sparse disposal locations lead to unsustainable practices such as street dumping and burning (Joshi et al., 2019). Tighter control of the use phase is important especially to stem the use of single-use plastics. A common strategy across the continent is the ban on plastic bags which has been implemented by 36 African countries (Attafuah-Wadee and Tilkanen, 2020). Furthermore, reuse business models can keep packaging in use for more than one cycle. The packaging is either returned to the business or retailer and can be refilled by the customer.

2.4 End of life

As previously said, plastic mismanagement at the end-of-life stage, i.e., when a plastic product becomes waste, is one of the most difficult environmental concerns and has created a hotspot in the plastic value chain (Oyinlola et al., 2022). Jambeck et al. (2015) estimated that mismanaged plastics in Africa can be as high as 85% although accurate data is not available. This low rate is as a result of a combination of factors. Kolade et al. (2022) suggest that environmental concerns are usually not a priority as most of the population are still struggling to meet the necessities of life, such as food and shelter. Furthermore, it has been widely reported that plastic recycling is not always economically viable, especially in Africa (Kreiger et al., 2014; Santander et al., 2020) as the costs of virgin plastics are usually cheaper than recycled plastics.

Another reason for this is the low number of recycling facilities and low volumes of available recycled plastics. The number of registered plastic recycling facilities in Africa is low. At the time of writing, there are only 89 materials recovery facilities (MRFs) and 68 recycling facilities based in Africa that are listed in the ENF directory of recycling companies (ENF, 2022). The three countries South Africa, Morocco and Nigeria together account for the largest shares (see Table 4.2). Most countries do only have one or none officially registered MRF and recycling facilities. Data on the capacity, annual processed volumes and types of technologies used in the facilities is lacking.

Furthermore, centralised municipal waste management systems are weak or non-existent across the continent. According to World Bank data, less than half of Africa's waste is collected formally, and systems for collection are usually non-existent in rural regions (Kaza et al., 2018). A significant amount of collection and recycling activities are semi-informal or informal which are unlikely to be

TABLE 4.2 Materials recovery facilities (MRFs) and recycling facilities based in Africa

Country	No. of registered MRFs	No. of registered recycling facilities
South Africa	49	20
Morocco	10	7
Nigeria	9	10
Ghana	4	7
Egypt	2	6
Tanzania	2	0
Tunisia	2	2
Mauritius	2	0
Namibia	2	0
Zimbabwe	1	3
Mozambique	1	1
Kenya	0	3
Algeria	1	3
Uganda	0	1
Ethiopia	0	1

(*Source*: ENF, 2022)

registered. These semi-formal recycling activities are characterised by suboptimal equipment and technologies and are typically driven by the informal sector (Oyinlola et al., 2022). Although this sector has many difficulties, the informal waste pickers are extremely conversant with the local environment and are often highly skilled at identifying and collecting valuable waste (Schröder et al., 2019). It has been suggested that compared to the formal sector, the informal sector can be more efficient in collecting and processing waste in the global south (Alhanaqtah, 2018). Therefore, it is important to integrate the informal sector in future developments of the waste sector (Wilson et al., 2006). In fact, without inclusion of the informal sector and improving the working conditions, equipment and operations, it will be difficult to achieve a circular plastic economy in Africa.

Over the past decade, several small-scale enterprises, most of which work actively in cooperation with the informal sector, have sprung up in several African countries. They are tackling the challenge by using plastic waste as an economic resource (Oyinlola et al., 2022). These small- and medium-sized enterprise (SMEs) have attracted growing support in the waste management value chain as it creates opportunities for collaboration to support a social, economic and environmental challenge. These organisations have increasingly received support from key actors such as local and foreign governments, investors, donor organisations, multinational companies, among others, as well as partnered with other actors in the value chains, e.g., the collection and disposal sector and recyclers, to facilitate sustainable waste management of plastics (Lane, 2018).

3 Digital technologies for improving circularity across the plastics value chain

Digital technologies have resulted in leapfrogging of several sectors in Africa such as finance (Kingiri and Fu, 2019), off-grid renewable energy (Annunziata et al., 2015), education (Oke and Fernandes, 2020) and agriculture (Syngenta, 2019). There are opportunities for digital technologies to address the plastic leakage and pollution issue (Jambeck et al., 2018). Scholars such as Chidepatil et al. (2020), Kolade et al. (2022), Mdukaza et al. (2018) and Oyinlola et al. (2022) have shown that with the right policies in place (Schroeder et al., 2023), digital technologies such as mobile applications, geographical information system (GIS) and artificial intelligence (AI) can play a significant role in a circular economy for plastics in Africa (Kolade et al., 2022; Oyinlola et al., 2022) and bridge the circularity divide between the global west and the global south (Barrie et al., 2022). This section discusses the use of digital technologies to eliminate leakage in the African plastic value chain.

The production phase is known to have many harmful substances; these can be identified and phased out with the use of AI. Similarly, AI could be used for optimising production which will result in a lower environmental footprint as well as quality control. Also, with the EPR scheme becoming more relevant, producers need to ensure traceability and accountability. Therefore, blockchain, which can be used to develop a more transparent and accountable system, can be employed to tag and track a product through its life cycle without fear of being altered. This technology is especially important given the inadequate infrastructure for waste management to track waste flows. Furthermore, the use of internet of things (IoT) devices coupled with mobile technology could be used to tackle the lack of data prevalent in the plastic value chain.

Oyinlola et al. (2022) highlighted that consumer behaviour is a major challenge on the African continent. One of the unfortunate outcomes of this is that currently numerous collectors are not able to get enough feedstock for recycling to be profitable. Digital tools could be used to encourage environmentally friendly habits through gamification (Hsu and Chen, 2021) via mobile apps. This could cover areas such as creating awareness and tracking of individual use of plastics, providing alternate to plastics, facilitating alternate delivery mechanisms instead of single use and providing a platform for a sharing economy. These would serve as incentives to drive a cultural and behavioural change as well as address issues such as level of literacy, environmental awareness and digitisation acceptance. Augmented reality/virtual reality (AR/VR) is another technology that could be leveraged for awareness, sensitisation and training. Similarly, the use of 5G and IoT sensors could support real-time communication between consumers and collectors.

Digital tools and technologies can be used to fill the gap of inadequate waste collection and management infrastructure. Web and mobile applications have been used to enhance the activities of the informal sector. For example, mobile applications underpinned with GIS have been developed to optimise the route

FIGURE 4.2 Digital technologies for the value chain

of informal waste collectors as well as incentivising them through schemes such as digital points that could be converted into mobile data credit. Mobile apps can serve as an essential interface for different actors in the plastic value chain to interact and communicate. The use of mobile apps across the continent is supported by the fact that there has been a surge in the number of smartphones over the past two decades (GSMA, 2020). AR/VR is another technology that could be used for building capacity on best practices for collection and sorting. As several SMEs embrace digitalisation, serverless computing or function as a service (FaaS) is a tool that could be adopted to eliminate the cost of infrastructural setup and deployment. Sorting is another task that can benefit from digitalisation, for example, robotics coupled with AI could support automation of the sorting process with minimal error from low-skilled workers.

Additive manufacturing, also known as three-dimensional (3D) printing, is another technology that can prevent leakage from the value chain by promoting upcycling of plastic waste (Oyinlola et al., 2023b). This technology allows plastic waste to be used as a feedstock for producing complex parts in remote areas while reducing the environmental footprint associated with traditional supply chain logistics (Kreiger and Pearce, 2013; Zhong and Pearce, 2018). This technology provides the opportunity to add significant value to the waste stream, thus incentivising consumers (Adefila et al., 2020; Oyinlola et al., 2018).

Figure 4.2 summarises various digital technologies that can be used to prevent leakage from the value chain.

4 Conclusion

There is a large discrepancy between the amount of plastic that is and will be produced in Africa and the existing collection, recovery and recycling infrastructure. Even though plastic production and use in Africa is currently low compared to other regions, given the anticipated increase in plastic production and consumption, it is important to focus on circular solutions to reduce current and future impacts of plastics. Furthermore, much of the current collection and recycling of plastic waste is carried out by the informal sector. Inclusion of the informal sector and improving the working conditions, equipment and operations will be an important element in achieving a circular plastic economy in Africa. Digital solutions can play an important role in enabling new business models that generate value from plastic waste.

Acknowledgement

This work was partly supported by the United Kingdom Research and Innovation (UKRI) Global Challenges Research Fund (GCRF) under Grant EP/T029846/1.

References

Adefila, A., Abuzeinab, A., Whitehead, T., Oyinlola, M., 2020. Bottle house: utilising appreciative inquiry to develop a user acceptance model. *Built Environ. Proj. Asset Manag.* 10, 567–583. https://doi.org/10.1108/BEPAM-08-2019-0072

Alhanaqtah, V., 2018. *Integrating the Informal Sector for Improved Waste Management in Rural Communities*, pp. 208–224. IGI Global. https://doi.org/10.4018/978-1-5225-7158-2.ch012

Annunziata, M., Bell, G., Buch, R., Patel, S., 2015. *Powering the Future: Leading the Digital Transformation of the Power Industry.* Boston, MA: General Electric Company.

Attafuah-Wadee, K., Tilkanen, J., 2020. Policy approaches for accelerating the circular economy in Africa [WWW Document]. Circ. Chatham House. https://circularecon omy.earth/publications/accelerating-the-circular-economy-transition-in-africa-pol icy-challenges-and-opportunities (accessed 12.12.20).

Ayeleru, O.O., Dlova, S., Akinribide, O.J., Ntuli, F., Kupolati, W.K., Marina, P.F., Blencowe, A., Olubambi, P.A., 2020. Challenges of plastic waste generation and management in sub-Saharan Africa: A review. *Waste Manag.* 110, 24–42.

Babayemi, J.O., Nnorom, I.C., Osibanjo, O., Weber, R., 2019. Ensuring sustainability in plastics use in Africa: consumption, waste generation, and projections. *Environ. Sci. Eur.* 31, 1–20.

Barrie, J., Anantharaman, M., Oyinlola, M., Schröder, P., 2022. The circularity divide: What is it? And how do we avoid it? *Resour. Conserv. Recycl.* 180, 106208. https://doi.org/10.1016/J.RESCONREC.2022.106208

Break Free From Plastic, 2019. BRANDED in search of the world's top corporate plastic polluters, Acesso em.15(11).

Chidepatil, A., Bindra, P., Kulkarni, D., Qazi, M., Kshirsagar, M., Sankaran, K., 2020. From trash to cash: how blockchain and multi-sensor-driven artificial intelligence can transform circular economy of plastic waste? *Adm. Sci.* 10, 23.

CircularEconomy.Earth, 2022. Trade | circulareconomy.earth | Chatham House [WWW Document]. *Chatham House Circ. Econ. Earth.* https://circulareconomy.earth/trade?year=2020&importer=ssf&category=36&units=weight&autozoom=1 (accessed 10.29.22).

D&B Hoovers, 2022. Plastics Product Manufacturing Companies [WWW Document]. URL https://www.dnb.com/business-directory/industry-analysis.plastics_product_manufacturing.html (accessed 11.16.22).

ENF, 2022. Plastic recycling plants in Africa – ENF recycling directory [WWW Document]. www.enfrecycling.com/directory/plastic-plant/Africa (accessed 10.18.22).

Euromap, 2020. Plastics resin production and consumption in 63 countries worldwide.

Geyer, R., Jambeck, J.R., Law, K.L., 2017. Production, use, and fate of all plastics ever made. *Sci. Adv.* 3, e1700782. https://doi.org/10.1126/sciadv.1700782

GSMA, 2020. The Mobile Economy. https://www.gsma.com/mobileeconomy/# (Accessed 14 May 2021).

Hsu, C.L., Chen, M.C., 2021. Advocating recycling and encouraging environmentally friendly habits through gamification: An empirical investigation. *Technol. Soc.* 66, 101621. https://doi.org/10.1016/J.TECHSOC.2021.101621

IISD, 2019. Companies Launch African Plastics Recycling Alliance [WWW Document]. SDG Knowl. HUB. https://sdg.iisd.org/news/companies-launch-african-plastics-recycling-alliance/ Accessed 14 May 2021)..

INTERPOL, 2020. INTERPOL report alerts to sharp rise in plastic waste crime [WWW Document]. www.interpol.int/en/News-and-Events/News/2020/INTERPOL-report-alerts-to-sharp-rise-in-plastic-waste-crime (accessed 11.16.22).

Jambeck, J.R., Geyer, R., Wilcox, C., Siegler, T.R., Perryman, M., Andrady, A., Narayan, R., Law, K.L., 2015. Plastic waste inputs from land into the ocean. *Science (80-.).* 347, 768–771.

Jambeck, J., Hardesty, B.D., Brooks, A.L., Friend, T., Teleki, K., Fabres, J., Beaudoin, Y., Bamba, A., Francis, J., Ribbink, A.J., Baleta, T., Bouwman, H., Knox, J., Wilcox, C., 2018. Challenges and emerging solutions to the land-based plastic waste issue in Africa. *Mar. Policy* 96, 256–263. https://doi.org/10.1016/j.marpol.2017.10.041

Joshi, C., Seay, J., Banadda, N., 2019. A perspective on a locally managed decentralized circular economy for waste plastic in developing countries. *Environ. Prog. Sustain. Energy* 38, 3–11.

Kaseva, M.E., Mbuligwe, S.E., 2005. Appraisal of solid waste collection following private sector involvement in Dar es Salaam city, Tanzania. *Habitat Int.* 29, 353–366.

Kaza, S., Yao, L., Bhada-Tata, P., Van Woerden, F., 2018. *What a Waste 2.0: A Global Snapshot of Solid Waste Management to 2050.* World Bank Publications.

Kingiri, A.N., Fu, X., 2019. Understanding the diffusion and adoption of digital finance innovation in emerging economies: M-Pesa money mobile transfer service in Kenya. *Innov. Dev.* https://doi.org/10.1080/2157930X.2019.1570695

Kolade, O., Odumuyiwa, V., Abolfathi, S., Schröder, P., Wakunuma, K., Akanmu, I., Whitehead, T., Tijani, B., Oyinlola, M., 2022. Technology acceptance and readiness of stakeholders for transitioning to a circular plastic economy in Africa. *Technol. Forecast. Soc. Change* 183, 121954. https://doi.org/10.1016/J.TECHFORE.2022.121954

Kreiger, M., Pearce, J.M., 2013. Environmental impacts of distributed manufacturing from 3-D printing of polymer components and products. *MRS Online Proc. Libr.* 1492, 85–90. https://doi.org/10.1557/opl.2013.319

Kreiger, M.A., Mulder, M.L., Glover, A.G., Pearce, J.M., 2014. Life cycle analysis of distributed recycling of post-consumer high density polyethylene for 3-D printing filament. *J. Clean. Prod.* 70, 90–96. https://doi.org/10.1016/J.JCLEPRO.2014.02.009

Lane, W., 2018. Oceans of plastics: developing effective African policy responses. South African Institute of International Affairs, 2018. www.jstor.org/stable/resrep29526

Lebreton, L., Andrady, A., 2019. Future scenarios of global plastic waste generation and disposal. *Palgrave Commun.* 5, 6. https://doi.org/10.1057/s41599-018-0212-7

Lerner, S., 2020. Africa's exploding plastic nightmare. *Intercept* 19. https://theintercept.com/2020/04/19/africa-plastic-waste-kenya-ethiopia/ (Accessed 20 May 2021).

Mdukaza, S., Isong, B., Dladlu, N., Abu-Mahfouz, A.M., 2018. Analysis of IoT-Enabled Solutions in Smart Waste Management, in: IECON 2018 – 44th Annual Conference of the IEEE Industrial Electronics Society. pp. 4639–4644. https://doi.org/10.1109/IECON.2018.8591236

Mofo, L., 2020. Future-proofing the plastics value chain in Southern Africa. WIDER Working Paper.

Narancic, T., O'Connor, K.E., 2019. Plastic waste as a global challenge: Are biodegradable plastics the answer to the plastic waste problem? *Microbiology* 165, 129–137.

OECD, 2022. Plastics use projections to 2060, in: Global Plastics Outlook: Policy Scenarios to 2060. OECD. https://doi.org/10.1787/AA1EDF33-EN

Oke, A., Fernandes, F.A.P., 2020. Innovations in teaching and learning: Exploring the perceptions of the education sector on the 4th industrial revolution (4IR). *J. Open Innov. Technol. Mark. Complex.* 6, 31. https://doi.org/10.3390/joitmc6020031

Oyinlola, M., Kolade, O., Schroder, P., Odumuyiwa, V., Rawn, B., Wakunuma, K., Sharifi, S., Lendelvo, S., Akanmu, I., Mtonga, R., Tijani, B., Whitehead, T., Brighty, G., Abolfathi, S., 2022. A socio-technical perspective on transitioning to a circular plastic economy in Africa. *SSRN Electron. J.* https://doi.org/10.2139/ssrn.4332904

Oyinlola, M., Okoya, S.A., Whitehead, T., Evans, M., Lowe, A.S., 2023b. The potential of converting plastic waste to 3D printed products in Sub-Saharan Africa. *Resour. Conserv. Recycl. Adv.* 17, 200129. https://doi.org/10.1016/j.rcradv.2023.200129

Oyinlola, M., Schröder, P., Whitehead, T., Kolade, S., Wakunuma, K., Sharifi, S., Rawn, B., Odumuyiwa, V., Lendelvo, S., Brighty, G., Tijani, B., Jaiyeola, T., Lindunda, L., Mtonga, R., Abolfathi, S., 2022. Digital innovations for transitioning to circular plastic value chains in Africa. *Africa J. Manag.* 8, 83–108. https://doi.org/10.1080/23322373.2021.1999750

Oyinlola, M., Whitehead, T., Abuzeinab, A., Adefila, A., Akinola, Y., Anafi, F., Farukh, F., Jegede, O., Kandan, K., Kim, B., Mosugu, E., 2018. Bottle house: A case study of transdisciplinary research for tackling global challenges. *Habitat Int.* 79, 18–29. https://doi.org/10.1016/j.habitatint.2018.07.007

Pacini, H., Shi, G., Sanches-Pereira, A., da Silva Filho, A.C., 2021. Network analysis of international trade in plastic scrap. *Sustain. Prod. Consum.* 27, 203–216.

Santander, P., Cruz Sanchez, F.A., Boudaoud, H., Camargo, M., 2020. Closed loop supply chain network for local and distributed plastic recycling for 3D printing: a MILP-based optimization approach. *Resour. Conserv. Recycl.* 154, 104531. https://doi.org/10.1016/J.RESCONREC.2019.104531

Schröder, P., Anantharaman, M., Anggraeni, K., Foxon, T.J., 2019. The Circular Economy and the Global South. In P. Schröder, M. Anantharaman, K. Anggraeni, T.J. Foxon (eds) *The Circular Economy and the Global South.* Routledge. https://doi.org/10.4324/9780429434006

Schroeder, P., Oyinlola, M., Barrie, J., Bonmwa, F., Abolfathi, S., 2023. Making policy work for Africa's circular plastics economy. *Resour. Conserv. Recycl.* 190, 106868. https://doi.org/10.1016/j.resconrec.2023.106868

SEI, UoY, 2020. Manufacturing pollution in sub-Saharan Africa and South Asia: Implications for the environment, health and future work: Main report. UNCTAD.

Statista, 2022. Plastic materials production by world region 2020 | Statista [WWW Document]. www.statista.com/statistics/281126/global-plastics-production-share-of-various-countries-and-regions/ (accessed 11.16.22).

Syngenta, 2019. How can digital solutions help to feed a growing world?

Tabuchi, H., Corkery, M., Mureithi, C., 2020. Big oil is in trouble. its plan: flood Africa with plastic. New York Times 30. https://www.nytimes.com/2020/08/30/climate/oil-kenya-africa-plastics-trade.html

UNEP, 2018. *Africa Waste Management Outlook.* Nairobi.

Wilson, D.C., Velis, C., Cheeseman, C., 2006. Role of informal sector recycling in waste management in developing countries. *Habitat Int.* 30, 797–808. https://doi.org/10.1016/j.habitatint.2005.09.005

Williams, M., Gower, R., Green, J., Whitebread, E., Lenkiewicz, Z., Schröder, P. 2019. No time to waste: Tackling the plastic pollution crisis before it's too late. Teddington: Tearfund.

Zhong, S., Pearce, J.M., 2018. Tightening the loop on the circular economy: Coupled distributedrecyclingandmanufacturingwithrecyclebotandRepRap3-Dprinting. *Resour. Conserv. Recycl.* 128, 48–58. https://doi.org/10.1016/J.RESCONREC.2017.09.023

5

DIGITAL INNOVATION ECOSYSTEM FOR THE CIRCULAR PLASTIC ECONOMY

Bosun Tijani, Muyiwa Oyinlola and Silifat Abimbola Okoya

1 Introduction

The word "innovation" has its origins from a Latin word "innovare" which means renewal (Barthwal, 2007). Various definitions of innovation are available, and innovation is multifaceted in nature (Alexander and Evgeniy, 2012). Innovation has been identified as an ongoing process of evolving and applying new knowledge and technologies to solve challenges and increase efficiency, affordability, reliability and sustainability (Ciriello et al., 2018). Innovation also serves as an important driver for economic growth and development in various facets of the economy such as education, commerce, transportation and telecommunication (Oluwatobi et al., 2015). Several countries have now placed innovation as a top priority on the political agenda (Blenker et al., 2006; Gerba, 2012) as it is viewed as an essential source of value creation (Fayolle, 2010) and leads to the development of a culture that can influence values and economic development (Carvalho et al., 2015).

Over the past decade, there has been a serious and growing interest in using digital transformation to scale some of the critical sectors of the economies across the continent. The rapid development of technology which allows for endless possibilities such as seamlessly connecting humans to humans, humans to machines and/or machines to machines as well as the ability to process data at an unprecedented scale has provided the ability to leapfrog development across all sectors. There are several examples in the literature of how businesses and nations have adopted technology to leapfrog development across the continent. Therefore, even though Africa has always been slow to catch up on global developments, the continent has experienced significant progress in the adoption of digital technologies across several sectors.

DOI: 10.4324/9781003278443-6

Digital transformation can be seen across a multitude of sectors such as finance (Kingiri and Fu, 2019), energy (Annunziata et al., 2015), education (Oke and Fernandes, 2020), water services (Amankwaa et al., 2021), agriculture (Syngenta, 2019) and the circular economy (Oyinlola et al., 2022b). The diffusion of innovations is therefore essential to accelerate the transition to a circular plastic economy (CPE) in Africa. This chapter provides insights into the dynamics that shape the development of the digital innovation ecosystem for the CPE and the role of local and international actors in this process.

2 Innovation System in Africa

As highlighted previously, technology is beginning to drive significant progress across all sectors, but for that to happen, a strong innovation system is required. In other words, innovation is accelerated through an ecosystem – a complex network of interconnected actors who function either independently or collaboratively as a whole, while sharing similar ideologies (Eggink, 2011; Pilinkienė and Mačiulis, 2014). The partnership in an ecosystem expands limitations in knowledge beyond one actor to allow for innovation with others (Adner, 2006). Innovation systems help build capacity across nations and sectors, and the level of maturity of an innovation system within a society drives the innovation outcomes and technical efficiency across its economy and sectors. A national innovation system requires a strong combination of organisations and institutions coming together to help build the capacity and boost the ability for innovation to happen in any society. Institutions such as universities, industrial partners, small-scale businesses, and investors all have a significant role to play in the innovation system.

However, despite historically weak innovation systems across Africa, there is an emerging innovation system that is driving significant value through digital transformation and strengthening technical efficiencies across African economies. This innovation system is defined by the shape and form in which knowledge and information are shared in today's connected world. This ecosystem is driving innovation in many sectors across the continent and causing the continent to leapfrog development across all the critical elements of what is expected in a typical innovation system. This innovation ecosystem is powered by two key things, firstly, the shape of knowledge and, secondly, technology innovation hubs, which can be considered the backbone of the ecosystem. This emerging innovation system in Africa differs from well-established economies in the Global West, where innovation is centralised in universities, research institutes and industries. This ecosystem has already resulted in several start-ups contributing to development in Africa. Figure 5.1 shows some of the tech start-ups in agriculture, public health and logistics. These organisations are supporting the application of technology and creative ideas to solving problems that are critical and peculiar to the African continent without having to invest in capital-intensive infrastructure.

FIGURE 5.1 Some technology start-ups across Africa

The emerging innovation system on the African continent has six unique attributes, which differentiate it from traditional innovation systems. These are as follows:

- Places emphasis on the role of social systems over individuals
- Enables and strengthens the formation of networks
- Independent of territorial construct
- Does not assume the existence of institutions and infrastructure
- Shapes institutions and inspires the formation of new institutions and practices
- Lowers barriers to membership and growth dynamics

These attributes show that a network approach is ideal for developing innovation capacity in Africa. This implies that rather than waiting for the different critical elements of the traditional innovation systems, a network approach could be adopted to leapfrog the capacity to innovate to solve critical issues in society. This emerging approach of distributed innovation systems means opportunities and relationships can be unlocked through the application of networks. These networks bring together different capacities which result in stimulating innovation.

The success of this network approach is driven by the fact that knowledge in today's world is located in a web of relationships, unlike previously when knowledge was concentrated and exclusive to places like unique academic institutions. The rapid proliferation of information and knowledge implies innovation is now more open and readily available. This is aided by the fact that being part of certain networks opens access to vast resources, knowledge and modern technology.

This network approach on the African continent will result in strengthening the innovation ecosystem, providing the opportunity and path to collective prosperity, accessing knowledge that can leapfrog opportunities easily available and providing

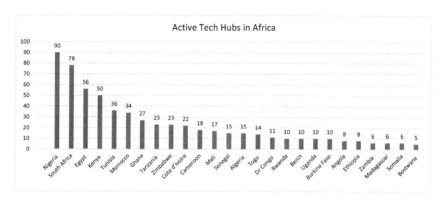

FIGURE 5.2   Tech hubs in Africa (Briter Bridges, 2019)

a mutually beneficial platform for relevant stakeholders. Furthermore, the network approach would help to eliminate barriers to innovation while allowing members to participate in new thinking. This thinking is likely to shape the way innovation capacity is developed across the continent.

Several factors are catalysing the growth of the African digital innovation ecosystem. Firstly, technology hubs, which are the backbone of this ecosystem, have sprung up across the continent within the last decade (Atiase et al., 2020). Briter Bridges (2019) reported that since the first hub, ihub, was founded in 2011, the number of technology hubs across the continent increased rapidly to 618 in 2019, as shown in Figure 5.2. These hubs which are driving and encouraging innovation have provided young people the opportunity to immerse themselves in technologies that result in innovations supporting development. The mushrooming of these tech hubs, which offer space and technology support for budding digital entrepreneurs, is empowering young Africans to be more creative and more innovative in their use of digital innovations. According to GSMA (Giuliani and Ajadi, 2019), the tech hubs offer support as incubators, accelerators, university-based innovation hubs, maker spaces, technology parks and co-working spaces. The tech hubs are instrumental in building digital innovation start-ups and a robust digital ecosystem where entrepreneurs can learn from as well as share ideas with like-minded innovators. Furthermore, tech hubs offer much-needed reliable internet access and electricity (Giuliani and Ajadi, 2019).

Secondly, increased access to capital has played a significant role in the growth of the ecosystem. Figure 5.3a shows that in 2022, while other world regions experienced a year-on-year decrease of up to 66%, Africa experienced a year-on-year growth of up to 171% (AVCA, 2022). Figure 5.3b shows that there has been an exponential increase in the volume of venture capital deals in Africa over the past 5 years, with 2022 predicted to have about 900 deals compared with 650 in 2021 and 319 in 2020. Figure 5.3c shows a similar exponential trend in the value of investment in African start-ups. This was US$0.3 bn in 2017 and US$5.2 bn in 2021 with 2022 predicted to be US$7 bn.

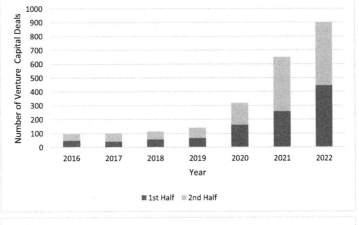

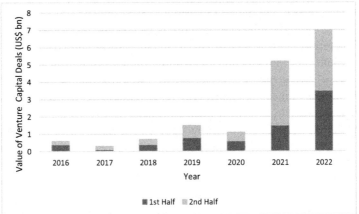

FIGURE 5.3 (a) The global venture capital market, (b) annual evolution of venture capital deals by volume and (c) annual evolution of venture capital deals by value source (AVCA, 2022)

Thirdly, the African Union Commission recognised the potential of digital innovations to create jobs, address poverty, reduce inequality and contribute to the sustainable development goals. In response, the commission developed a comprehensive digital transformation strategy for Africa, setting several specific targets for 2030 (African Union, 2020). This enabling policy is targeting core challenge areas in scaling up education and lack of infrastructure. Similarly, different initiatives have been implemented across Africa to leverage digital accessibility and affordability (DO4Africa, 2022). Entities such as schools, institutions and non-governmental organisations (NGOs) are pushing hard to break the digital divide in Africa and increase digital literacy in the population for capacity building (DO4Africa, 2022). These enabling policies across the continent are also catalysing the growth of the emerging innovation ecosystem. Fourthly, Africa's digital infrastructure has been rapidly evolving in recent times; for example, Africa now has the fastest-growing internet penetration rate in the world (Granguillhome Ochoa et al., 2022; GSMA, 2020), and the continent has attracted significant investment in digital platforms such as the Google AI hub in Ghana and Facebook hub in Kenya. Also, economic growth in sub-Saharan Africa has been at a record pace with countries recording some of the fastest growth rates globally over the past two decades (Fuje and Yao, 2022). This growth has in turn contributed to the evolution of the digital innovation ecosystem. Other catalysts include the proliferation of smartphones on the continent (GSMA, 2020) and a young demographic profile, with almost 60% of the population under 25 (Statista, 2021). This bourgeoning youth population is deeply connected to knowledge and is now building a significant foundation for innovation to thrive.

3 Innovation in Africa's CPE

The sectoral systems of innovation framework (Malerba, 2002) was adopted to examine the factors that affect innovation in the CPE. This framework ensures that the mapping and diagnostic exercise takes all relevant stakeholders into consideration and understands the dynamics of the interactions between different groups of stakeholders and how effective linkages can be developed in the sector to achieve expected outcomes. Malerba (2005) highlighted that sectoral systems are based on the three building blocks discussed below.

3.1 Actors and Networks

Innovation within a sector is a process that involves systematic interactions among a wide variety of actors for the generation and exchange of knowledge relevant to innovation and its commercialisation (Dahesh et al., 2020). This interaction is facilitated and accelerated by networks. Oyinlola et al. (2022) suggested that actors in the circular plastic innovation ecosystem can be classified into the following stakeholder groups: (1) digital innovation firms/start-ups, (2) civil society, (3) governments/policymakers, (4) waste management organisations, (5) academia, (6) investors and (7) community.

An increasing number of actors are presenting as start-ups going out of their way to create new ideas, approaches and interventions for plastic waste management in Africa. These actors are creating social, economic and environmental value. Some of these include Wecyclers, Capture Solutions, Mr. Green Africa, Pakam and Yo-Waste. A comprehensive list of these start-up actors was presented by Oyinlola et al. (2022b). Start-ups that employ technology are increasing in value and volume as AVCA (2022) reported that Cleantech (companies that harness or develop technology that seeks to improve environmental sustainability or to reduce the negative environmental impact of natural resources consumed through human activities) "rose five places to become the second most active vertical among technology or tech-enabled companies that successfully raised venture capital in the first half of 2022" (AVCA, 2022). The African Private Equity and Venture Capital Association (AVCA, 2022) predicts that investment in Cleantech will continue to increase in volume and value as "impact investors motivated to meet Africa's sustainable development agenda back the growing number of African entrepreneurs delivering innovative, effective, and sustainable solutions to pressing socio-environmental challenges".

On the other hand, there are several networks in place fostering interactions between actors in Africa's circular economy, examples of these include the African Circular Economy Network (ACEN), Circular Economy Network, Marine Plastic, Coastal Communities Protect Network and DITCh Plastic Network. More networks are being formed across the continent to drive the circular plastic agenda.

Scholars have highlighted the need for stakeholder collaboration to strengthen the innovation ecosystem. This is extremely vital as stakeholders of the CPE primarily currently operate in silos (Oyinlola et al., 2022). It has been reported that inadequate collaboration and coordination among different sets of stakeholders pose a significant challenge to the progress of the circular economy (Sarja et al., 2021). Multi-stakeholder synergy and collaboration, which can be achieved through networks, can invigorate the ecosystem, support new ideas and innovations and accelerate the diffusion of innovation across the continent. These collaborative initiatives need to be strategic and inclusive and underpinned by better communication and networking approaches. These networking and collaboration opportunities will enable an omnidirectional, heterarchical process of stakeholder engagement in the CPE (Obembe et al., 2021). This approach is, in turn, best suited to the co-creation of innovations and a higher level of the ongoing commitment from stakeholders. Diaz et al. (2021) noted that synergies across the board would be facilitated by considering circularity as a socio-technical challenge, while Nikas et al. (2022) suggested utilising a multidisciplinary approach where communities create knowledge jointly with non-scientific stakeholders such as civil societies, industries and policymakers. The ecosystem approach for the CPE in Africa will lead to increased collaboration and knowledge transfer among actors (Kruss and Visser, 2017).

3.2 Institutions

The actions and interactions of actors are influenced by what is referred to as institutions, which include norms, common habits, established practices, policies, laws and standards (Malerba, 2002). They may be binding or non-binding and formal or informal (e.g. patent laws or specific regulations vs. traditions and conventions). The role of institutions in fostering a circular economy is paramount to promoting entrepreneurial and innovative skills for sustainable development. These institutions include those that support regulation and policy as well as those involved in regional path development of the circular plastic innovation ecosystem. Institutional stakeholder theory suggests that organisations produce both positive and negative impacts on stakeholders who, in addition, compel them to change positively (Meherishi et al., 2019). It is therefore essential for these institutions to foster engagements that nurture collaborative relationships between stakeholders and co-create viable value for all. Related institutional organisations fostering the CPE in Africa include the African Development Bank (AfDB), United Nations Environment Programme (UNEP), Ellen MacArthur Foundation and Global Plastic Action Partnership (GPAP).

These institutions play a significant role in governing the interactions of stakeholders within the ecosystem. Reed et al. (2009) suggest that this can be effectively managed using the following steps: communicating and explaining the social and natural trends affected by the actions; identifying the organisations, groups, individuals and other stakeholders critical to the process; and prioritising them for involvement and decision-making. Furthermore, Williamson and De Meyer (2012) propose that developing a successful ecosystem requires a lead organisation that acts as the prime planner and guide to identify potential value creation. This implies that these related institutional organisations need to take leadership in driving the CPE. Williamson and De Meyer (2012) further highlight that the lead organisation has the primary responsibility of identifying and appealing to potential partners, identifying roles for partners, reducing risks and building confidence while developing an environment for co-learning.

To strengthen the innovation impact, institutions need to imbibe suitable considerations in the areas of resources, governance, strategy, leadership, organisational structure, human resource management, people, partners, technology and clustering (Durst and Poutanen, 2013). It is also important to incorporate an innovation policy as most models lack a thorough perspective because they primarily focus on input factors and capacity to innovate (Yawson, 2021). The policy should integrate core challenges (Holdren, 2008) and systematic approaches. The implementation process should also plan for some possible constraints such as limited infrastructure, awareness among actors, political support and procedures (Preston et al., 2019).

3.3 Knowledge and Technological Domains

Knowledge plays a critical role in the evolution of the CPE. Therefore, the primary stakeholders in the ecosystem must be subject matter experts in their various fields with the required technical and soft skills to add value. Formal and informal training also need to be provided (Kumar et al., 2021) to ensure an understanding of the process and engagement among stakeholders and the general public. The concept of resource management and interaction is also critical to establish an effective and efficient working ecosystem.

Malerba (2002) suggested that "one knowledge domain refers to the specific scientific and technological fields at the base of innovative activities in a sector" while the second knowledge domain relates to applications, users and demand for sectoral products. From the assertions of Oyinlola et al. (2022), it can be observed that the first knowledge domain is currently predominantly focused on "Recycling", while other significant areas of the circular economy, such as "Reduce" and "Reuse", are not yet as widespread in Africa. Oyinlola et al. (2022) further indicated that among the public across Africa, there is still a significant lack of education and awareness on sustainable waste management. Critical knowledge domains that need to be developed include changing consumer behaviours, socio-technical transition and alternate sustainable materials to plastic.

In terms of the technologies for the CPE, Kolade et al. (2022) examined the readiness and acceptance of stakeholders to ten frontier technologies and found that only three [i.e. mobile app, geographic information systems (GIS) and internet of things (IoT)] have appropriate technology readiness to be implemented in CPE strategies. They further highlighted that mobile apps are by far the most developed digital tool across Africa. Similarly, Oyinlola et al. (2022) identified mobile apps, blockchain and three-dimensional (3D) printing as critical technologies for niche CPE innovations for Africa. They highlight the need to develop capacity to deploy, develop, operate and maintain other emerging technologies across the continent.

4 Conclusion

Given the preceding discussion, the CPE in Africa can leverage the continent's emerging innovation ecosystem to drive progress. Developing a strong ecosystem will lead to increased collaboration and knowledge transfer among actors as well as alleviate some of the challenges by promoting interconnectivity of shared vision, values, information and resources. Figure 5.4 presents the circular plastic innovation ecosystem comprising actors and networks, knowledge and technical domains and institutions. A combination of these elements of the ecosystem will foster a culture of innovation as well as result in key players combining efforts to build capacity and strengthen the possibility of innovations happening in society. The synergies brought about by this collaboration will lead to technical efficiency which is an opportunity to leapfrog Africa's CPE.

Building a culture of innovation is paramount for these possibilities, and a budding ecosystem provides an encouraging environment for innovation to occur. An enabling ecosystem will help create, transform and communicate knowledge,

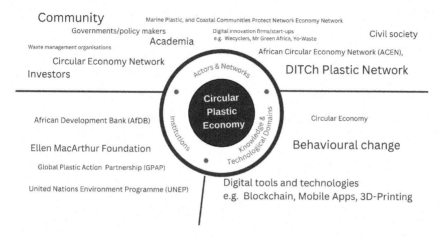

FIGURE 5.4 CPE innovation ecosystem

thereby nurturing innovation. Furthermore, building capacity will create jobs for young people especially as it is anticipated that the shift to a circular economy will redefine the nature of jobs and competencies (UNESCO, 2021).

Acknowledgement

This work was partly supported by the United Kingdom Research and Innovation (UKRI) Global Challenges Research Fund (GCRF) under Grant EP/T029846/1.

References

Adner, R., 2006. Match your innovation strategy to your innovation ecosystem. *Harv. Bus. Rev.* 84, 98.

African Union, 2020. Draft digital transformation strategy for Africa (2020–2030). https://archives.au.int/handle/123456789/8758 (Accessed 13 May 2021)

Alexander, U., Evgeniy, P., 2012. The entrepreneurial university in Russia: From idea to reality. *Procedia-Social Behav. Sci.* 52, 45–51.

Amankwaa, G., Heeks, R., Browne, A.L., 2021. Digital innovations and water services in cities of the global south: A systematic literature review. *Water Altern.* 14, 619–644. www.water-alternatives.org/index.php/alldoc/articles/vol14/v14issue2/637-a14-2-15

Annunziata, M., Bell, G., Buch, R., Patel, S., 2015. *Powering the Future: Leading the Digital Transformation of the Power Industry.* General Electric Company: Boston, MA.

Atiase, V.Y., Kolade, O., Liedong, T.A., 2020. The emergence and strategy of tech hubs in Africa: Implications for knowledge production and value creation. *Technol. Forecast. Soc. Change* 161, 120307. https://doi.org/10.1016/J.TECHFORE.2020.120307

AVCA, 2022. 2022 H1 African Venture Capital Activity Report. https://www.avca-africa.org/media/3064/02080-avca-vc-mid-year-report-sept22-online_2.pdf

Published by African Private Equity and Venture Capital Association (Accessed 01 November 2022)

Barthwal, R.R., 2007. *Industrial Economics: An Introductory Text Book*. New Age International.

Blenker, P., Dreisler, P., Kjeldsen, J., 2006. *Entrepreneurship Education: The New Challenge Facing the Universities*. Department of Management, Aarhus School of Business.

Briter Bridges, 2019. The backbone of Africa's tech ecosystem: 618 active tech hubs in Africa [WWW Document]. https://briterbridges.com/618-active-tech-hubs (accessed 11.19.22).

Carvalho, L., Costa, T., Mares, P., 2015. A success story in a partnership programme for entrepreneurship education: Outlook of students perceptions towards entrepreneurship. *Int. J. Manag. Educ.* 9, 444–465.

Ciriello, R.F., Richter, A., Schwabe, G., 2018. Digital innovation. *Bus. Inf. Syst. Eng.* 60, 563–569.

Dahesh, M.B., Tabarsa, G., Zandieh, M., Hamidizadeh, M., 2020. Reviewing the intellectual structure and evolution of the innovation systems approach: A social network analysis. *Technol. Soc.* 63, 101399.

Diaz, A., Schöggl, J.P., Reyes, T., Baumgartner, R.J., 2021. Sustainable product development in a circular economy: Implications for products, actors, decision-making support and lifecycle information management. *Sustain. Prod. Consum.* 26, 1031–1045. https://doi.org/10.1016/J.SPC.2020.12.044

DO4Africa, 2022. Digital literacy in Africa [WWW Document]. www.do4africa.org/en/digital-literacy-in-africa/ (accessed 11.19.22).

Durst, S., Poutanen, P., 2013. Success factors of innovation ecosystems-initial insights from a literature review. Co-create 2013, 27.

Eggink, M.E., 2011. *The Role of Innovation in Economic Development*. University of South Africa Pretoria.

Fayolle, A., 2010. Insights from an international perspective on entrepreneurship education. *Handb. Res. Entrep. Educ.* 3, 1–9.

Fuje, H., Yao, J., 2022. Africa's rapid economic growth hasn't fully closed income gaps [WWW Document]. IMF. www.imf.org/en/Blogs/Articles/2022/09/20/africas-rapid-economic-growth-hasnt-fully-closed-income-gaps (accessed 11.19.22).

Gerba, D.T., 2012. Impact of entrepreneurship education on entrepreneurial intentions of business and engineering students in Ethiopia. *African J. Econ. Manag. Stud.* 3(2), 258–277.

Giuliani, D., Ajadi, S., 2019. 618 active tech hubs: The backbone of Africa's tech ecosystem [WWW Document]. Mob. Dev. www.gsma.com/mobilefordevelopment/blog/618-active-tech-hubs-the-backbone-of-africas-tech-ecosystem/

Granguillhome Ochoa, R., Lach, S., Masaki, T., Rodríguez-Castelán, C., 2022. Mobile internet adoption in West Africa. *Technol. Soc.* 68, 101845. https://doi.org/10.1016/J.TECHSOC.2021.101845

GSMA, 2020. The Mobile Economy. https://www.gsma.com/mobileeconomy/# (Accessed 14 May 2021)

Holdren, J.P., 2008. Science and technology for sustainable well-being. *Science (80-.).* 319, 424–434.

Kingiri, A.N., Fu, X., 2019. Understanding the diffusion and adoption of digital finance innovation in emerging economies: M-Pesa money mobile transfer service in Kenya. *Innov. Dev.* https://doi.org/10.1080/2157930X.2019.1570695

Kolade, O., Odumuyiwa, V., Abolfathi, S., Schröder, P., Wakunuma, K., Akanmu, I., Whitehead, T., Tijani, B., Oyinlola, M., 2022. Technology acceptance and readiness of stakeholders for transitioning to a circular plastic economy in Africa. *Technol. Forecast. Soc. Change* 183, 121954. https://doi.org/10.1016/J.TECHFORE.2022.121954

Kruss, G., Visser, M., 2017. Putting university–industry interaction into perspective: A differentiated view from inside South African universities. *J. Technol. Transf.* 42, 884–908.

Kumar, Rakesh, Verma, A., Shome, A., Sinha, R., Sinha, S., Jha, P.K., Kumar, Ritesh, Kumar, P., Das, S., Sharma, P., 2021. Impacts of plastic pollution on ecosystem services, sustainable development goals, and need to focus on circular economy and policy interventions. *Sustainability* 13, 9963.

Malerba, F., 2002. Sectoral systems of innovation and production. *Res. Policy* 31, 247–264. https://doi.org/10.1016/S0048-7333(01)00139-1

Malerba, F., 2005. Sectoral systems of innovation: A framework for linking innovation to the knowledge base, structure and dynamics of sectors. *Econ. Innov. New Technol.* 14, 63–82. https://doi.org/10.1080/1043859042000228688

Meherishi, L., Narayana, S.A., Ranjani, K.S., 2019. Sustainable packaging for supply chain management in the circular economy: A review. *J. Clean. Prod.* 237, 117582.

Nikas, A., Xexakis, G., Koasidis, K., Acosta-Fernández, J., Arto, I., Calzadilla, A., Domenech, T., Gambhir, A., Giljum, S., Gonzalez-Eguino, M., Herbst, A., Ivanova, O., van Sluisveld, M.A.E., Van De Ven, D.J., Karamaneas, A., Doukas, H., 2022. Coupling circularity performance and climate action: From disciplinary silos to transdisciplinary modelling science. *Sustain. Prod. Consum.* 30, 269–277. https://doi. org/10.1016/J.SPC.2021.12.011

Obembe, D., Al Mansour, J., Kolade, O., 2021. Strategy communication and transition dynamics amongst managers: A public sector organization perspective. *Manag. Decis.* 59, 1954–1971. https://doi.org/10.1108/MD-11-2019-1589

Oke, A., Fernandes, F.A.P., 2020. Innovations in teaching and learning: Exploring the perceptions of the education sector on the 4th industrial revolution (4IR). *J. Open Innov. Technol. Mark. Complex.* 6, 31. https://doi.org/10.3390/joitmc6020031

Oluwatobi, S., Efobi, U., Olurinola, I., Alege, P., 2015. Innovation in Africa: Why institutions matter. *South African J. Econ.* 83, 390–410.

Oyinlola, M., Kolade, S., Odumuyiwa, V., Schröder, P., Whitehead, T., Wakunuma, K., Lendelvo, S., Rawn, B., Sharifi, S., Akanmu, I., Brighty, G., Mtonga, R., Tijani, B., Abolfathi, S., 2022. A socio-technical perspective on transitioning to a circular plastic economy in Africa. *SSRN Electron. J.* https://doi.org/10.2139/ssrn.4332904.

Oyinlola, M., Schröder, P., Whitehead, T., Kolade, S., Wakunuma, K., Sharifi, S., Rawn, B., Odumuyiwa, V., Lendelvo, S., Brighty, G., Tijani, B., Jaiyeola, T., Lindunda, L., Mtonga, R., Abolfathi, S., 2022b. Digital innovations for transitioning to circular plastic value chains in Africa. *Africa J. Manag.* https://doi.org/10.1080/23322 373.2021.1999750

Pilinkienė, V., Mačiulis, P., 2014. Comparison of different ecosystem analogies: The main economic determinants and levels of impact. *Procedia-social Behav. Sci.* 156, 365–370.

Preston, F., Lehne, J., Wellesley, L., 2019. An inclusive circular economy. *Priorities Dev. Ctries.* https://www.chathamhouse.org/sites/default/files/publications/research/2019-05-22-Circular%20Economy.pdf (Accessed 13 May 2021).

Reed, M.S., Graves, A., Dandy, N., Posthumus, H., Hubacek, K., Morris, J., Prell, C., Quinn, C.H., Stringer, L.C., 2009. Who's in and why? A typology of stakeholder analysis methods for natural resource management. *J. Environ. Manage.* 90, 1933–1949.

Sarja, M., Onkila, T., Mäkelä, M., 2021. A systematic literature review of the transition to the circular economy in business organizations: Obstacles, catalysts and ambivalences. *J. Clean. Prod.* 286, 125492. https://doi.org/10.1016/J.JCLEPRO.2020.125492

Statista, 2021. Africa: Population by age group 2020 [WWW Document]. Popul. Africa 2020, by age Gr. www.statista.com/statistics/1226211/population-of-africa-by-age-group/ (accessed 9.26.21).

Syngenta, 2019. How can digital solutions help to feed a growing world? www.syngentafoundation.org/file/12811/download (Accessed 13 May 2021).

UNESCO, 2021. Skills for the circular economy [WWW Document]. https://unevoc.unesco.org/home/Skills+for+the+circular+economy (accessed 8.17.22).

Williamson, P.J., De Meyer, A., 2012. Ecosystem advantage: How to successfully harness the power of partners. *Calif. Manage. Rev.* 55, 24–46.

Yawson, R.M., 2021. The ecological system of innovation: A new architectural framework for a functional evidence-based platform for science and innovation policy. arXiv Prepr. arXiv2106.15479.

PART II

Digitisation in Action

6

UTILISING PLASTIC WASTE TO CREATE 3D-PRINTED PRODUCTS IN SUB-SAHARAN AFRICA

Muyiwa Oyinlola, Silifat Abimbola Okoya and Timothy Whitehead

1   Introduction

Plastics offer a wide range of benefits such as affordability, low weight, flexibility and versatility (Mwanza et al., 2018). This makes them extremely useful globally, especially in the Global South where it is commonplace to have infrastructure deficits. However, in recent times, there has been increased awareness of the significant environmental and health impacts associated with plastic pollution (Ryberg et al., 2019; Thompson et al., 2009; Wabnitz and Nichols, 2010). The problem of plastic pollution is particularly exacerbated in Africa due to poor infrastructure and waste management systems which are suboptimal. A promising approach for tackling this challenge is the adoption of the circular plastic economy (Oyinlola et al., 2022b). The circular plastic economy is a system which employs the principles of the circular economy to the plastic value chain, including design, manufacture, use and end-of-life phase.

The concept of the circular economy has been explored by several scholars including Araujo Galvão et al. (2018), Berg et al. (2018), Gall et al. (2020) and Murray et al. (2017), and many have shown that digital technologies such as mobile applications, geographical information system (GIS) and three-dimensional (3D) printing, can play a significant role in the circular economy (and by extension, the circular plastic economy) in Africa (Kolade et al., 2022b; Oyinlola et al., 2022). Adopting digital technologies for the circular plastic economy can be revolutionary in terms of bridging the circularity divide (Barrie et al., 2022) as well as positively disrupting the landscape (Kolade et al., 2022b).

Digital technologies have been applied to varying extents across multiple sectors in Africa such as finance (Kingiri and Fu, 2019), energy (Annunziata et al., 2015), education (Oke and Fernandes, 2020), water services (Amankwaa et al.,

DOI: 10.4324/9781003278443-8

2021) and agriculture (Syngenta, 2019). The application of digital technology has resulted in leapfrogging in many areas, for example, up on till the 1990s, Africa's infrastructure for landline telephones was grossly inadequate and required huge investments to get a substantial number of the population connected. However, with the arrival of the global system for mobile communications (GSM), the communication sector has leapfrogged, cutting the need for heavy investment in landline infrastructure and improved access to mobile phones as it has been reported that "the world's poorest households are more likely to have a mobile phone than a toilet" (Devarajan, 2010).

This chapter makes a case for additive manufacturing, also known as 3D printing (a method of creating 3D solid items from a digital file) as one of the leading digital technologies for the circular plastic economy. Drawing on case studies in education, medicine, construction and local industries, this contribution illustrates how local plastic waste can be used to create new, innovative, locally made products which meet specific local needs.

2   3D printing as a promising intervention

A schematic of the process of converting plastic to 3D products is presented in Figure 6.1. First, the plastics are collected and sorted based on type to ensure each batch is homogeneous. The sorting is followed by cleaning, which involves removing labels and the label glue, washing and rinsing. This process ensures that

FIGURE 6.1   The basic steps of converting waste to 3D-printed products (Oyinlola et al., 2023)

the batch contains no contaminants. The cleaned homogeneous plastic batch will then be grinded into granules in readiness for extrusion to filaments. The granules must be dried as moisture content can affect the extrusion process. The granules can then be fed into the extruder through a hopper (Garmulewicz et al., 2016; Singh et al., 2017; Zander et al., 2018; Zhong and Pearce, 2018). The extruded filament is then cooled and spooled and can be used for 3D printing products. The production of a 3D-printed object is accomplished using additive processes. An object is built in an additive technique by laying down successive layers of material until the object is complete. Each of these levels is a thinly sliced cross-section of the object.

Additive manufacturing is widely recognised by international organisations such as the UK Foreign, Commonwealth and Development Office (FCDO), UNICEF and the United Nations as a leading frontier technology which would support international development (Ramalingam et al., 2016). Currently, the technology is at a "tipping point", where it is becoming a feasible manufacturing technique and is considered to be the cornerstone of the next industrial revolution (Rauch et al., 2016). This game-changing technology is expected to have substantial impact in low- and middle-income countries (LMICs) as the cost of an entry-level printer has declined from $30,000 to $200 in the last two decades (Berman, 2012; O'Connell and Haines, 2022). This is supported by the fact that in more recent times, 3D printers have been produced locally in Africa using e-waste such as servo motors from two-dimensional (2D) printers. For example, Kumar et al. (2021) repurposed e-waste to develop the essential components such as the stepper motors, power supply and iron supporting framework for a 3D printer–scanner hybrid from e-waste. Similarly, Simons et al. (2019) designed and developed a Delta 3D printer using salvaged e-waste materials. Three vertical axes spaced 120 degrees apart are used to move the printer. They used 17 stepper motors to operate the numerous carriages on the vertical axes to achieve accuracy and speed. Locally available square pipes, bearings and a 3D-printed rail were used in place of typical linear rails. A carriage support system was created, as well as a reasonably inexpensive but stable linear rail.

3D printing will allow communities to leapfrog traditional manufacturing (Kolade et al., 2022a; Swiss Business Hub and Swiss, 2018). Traditional manufacturing is characterised by manufacturing things in one place using highly capital-intensive methods such as injection moulding, in big factories. Products from these factories are then shipped through complex global supply chain networks which can be susceptible to delays when things go wrong such as the recent Suez canal blockage (Lawrence, 2021). Compared to traditional manufacturing methods, 3D printing provides a real opportunity to have a distributed supply chain where products or parts that people actually want or need in that local community can be made locally close to the point of demand. This means regardless of being in a rural village or a big city, products can be made locally, using local resources. Furthermore, this technology allows users to produce complex parts with essentially no waste, as it creates products layer by layer and can control the fill density of the product (Celik, 2020). Therefore, 3D printing offers the opportunity to reuse plastic materials in producing complex parts in remote areas while reducing the environmental footprint associated with traditional supply chain logistics. Converting the relative low-value plastic waste into a product that people can use makes the waste stream more of a resource than trash. This approach has become more popular as even the Dutch airline, KLM, started using polyethylene terephthalate (PET) bottles to make tools to repair and maintain its aircraft. According to the airline, empty bottles are collected at the end of every flight and transformed into filament, which is then used in a 3D printer to create new products (KLM, 2019).

With a promising growth forecast, 3D printing is likely to reduce the cost of manufacturing and result in shorter lead times while minimising the reliance on unsustainable and unreliable supply chains as well as create new businesses and support wealth generation (Shah et al., 2019). This is very significant for small and medium enterprises (SMEs) as it lowers the barriers to manufacturing since there are no tooling costs and one printer can produce specific parts for different applications at the same time. Chong et al. (2018) reviewed different initiatives in the area of 3D-assisted hybrid manufacturing and concluded that 3D manufacturing is promoted by major industrial countries as a technology starting point for the future of manufacturing.

Implementing 3D printing has the potential to increase recycling rate in Africa due to being a distributed recycling approach. Various scholars (Baechler et al., 2013; Cruz Sanchez et al., 2017; Kreiger et al., 2014; Woern et al., 2018) have shown that using waste materials for 3D printing allows consumers to recycle waste in their community which is a suitable intervention for improving recycling rates in LMICs. This decentralised approach differs from the usual centralised approach in the Global West which involves the transportation of the high-volume and low-weight polymers (Kreiger et al., 2014; Santander et al., 2020). Given the lack of adequate waste collection, transportation and recycling infrastructure, the concept of decentralised recycling is better suited for Africa. A centralised approach is not always economically viable due to the wide geographical spread in Africa (Kreiger et al., 2014; Santander et al., 2020) and can have a significant environmental footprint (Ragaert et al., 2017) due to the greenhouse gas emissions associated with the collection and transportation of the waste materials (Garmulewicz et al., 2016). Life cycle analysis of the distributed recycling method has shown that it has less embodied energy compared with the best-case scenario for centralised recycling. For example, Kreiger et al. (2014) note that more than 100 million MJ of energy was conserved annually, along with substantial reductions in greenhouse gas emissions. This is mainly because the process eliminates the environmental footprint associated with transporting waste to centralised collection points (Garmulewicz et al., 2016). In addition, 3D printing can increase the value of waste plastic by up to 20 times (Oyinlola et al., 2023). This can go a long way in incentivising sustainable plastic practices in communities (Adefila et al., 2020).

The rapid growth in the 3D printing sector and increased growth of open-source designs have fostered the use of 3D printers across Africa; however, the high cost of filaments, which are usually imported, remains a prohibiting factor. However, combining the prospects of repurposing electronic waste to develop 3D printers which will utilise filaments made from waste plastics implies that this technology aligns with several principles of the circular economy (Pavlo et al., 2018) such as supporting decentralised recycling, upcycling, reuse and distributed manufacturing (Sanchez et al., 2020). Furthermore, the application of 3D printing in Africa will lead to a quadruple bottom-line effect by increasing value (profit), reducing waste (planet), encouraging social well-being (people) and generating technical innovation (progress) (Oyinlola et al., 2023).

However, it is pertinent to note that the technology is still relatively in its infancy, therefore much more needs to be done to accelerate its adoption such as infrastructure (Oyinlola et al., 2023) and policy landscape (Schroeder et al., 2023). Furthermore, 3D printing is not capable of high-volume manufacturing, therefore much more suited to bespoke one-off production such as small-scale production, higher value items and replacement parts that are not readily available (Kolade et al., 2022a).

3 Case studies of products

There is a growing need for sustainably produced products by consumers (Cruz Sanchez et al., 2017; Feeley et al., 2014). Implementing 3D printing offers endless transformational possibilities including the creation of new, innovative, locally made products which meet specific local needs (Savonen, 2019). This section reviews how 3D printing has been utilised to aid development using various case studies. It specifically focuses on the application of 3D printing in four critical sectors: education, medicine, construction and local industries.

3.1 Education

Education underpins sustainable development; therefore, one way of tackling the developmental divide is by delivering quality education (Duran and Parker, 2021). Quality education will build the capacity and capability of local populations to tackle developmental challenges. 3D printing in education transcends numerous and separate clusters of works (Ford and Minshall, 2019), and it has been used to improve the quality of education across all levels ranging from primary to tertiary education. Schelly et al. (2015) investigated the use of open-source 3D printing technologies for education and concluded that science, technology, engineering and mathematics (STEM) education and career and technical education (CTE) can be improved via 3D printing because it provides a sense of empowerment resultant from active participation, as well as fosters cross-curriculum engagement. Obwoge et al. (2018) noted that 3D printing could be used for developing teaching

FIGURE 6.2 A 3D-printed microscope using waste plastics

and learning models to cater to different learning styles, teaching of 3D design in engineering education and for producing simple laboratory equipment.

Owen (2018) and Rogge et al. (2017) reported that the deployment of low-cost 3D-printed microscopes (see Figure 6.2) in Kenyan schools has been transformational in terms of improving access to teaching and learning equipment. Similarly, Garcia et al. (2018) reported that the fast development of 3D printing has created a novel learning and teaching approach required for medical education. Similarly, Del Rosario et al. (2022) noted that 3D printing has a unique impact on the medical education field by facilitating the manufacture of teaching equipment and aids such as the highly multifaceted robotic microscope, OpenFlexure. In architecture, 3D printing allows architects to produce complex parametric modelling geometries (Paio et al., 2012), while in engineering, Crowe et al. (2021) noted that 3D printing was effective in teaching about water interactions on both hydrophilic and hydrophobic surfaces. Pinger et al. (2019) reviewed the application of 3D printing in chemistry classrooms and concluded that the use of 3D-printed models for improved visualisation of chemical phenomena, as well as the educational use of 3D-printed laboratory devices, improves chemistry education. Santos et al. (2019) concluded that "3D printing with girls will be most successful when the context factor of role models, the child factor of engaging and relevant experiences, and the context factor of free play are taken into account".

Despite the numerous benefits of 3D printing in the education sector, Berman et al. (2018) who explored the 3D printing process for young children in curriculum-aligned making in the classroom noted that some students appeared more interested in printing designs for their visual aesthetics instead of their significance for their presentations, while others had difficulties in gauging the designs which begs the need for simplification of the process (Berman et al., 2018).

These examples illustrate that 3D printing can be utilised to transform teaching and learning in sub-Saharan Africa by giving teachers and students unprecedented access to aids. In practice, a financially challenged school in a remote location with a 3D printer and internet connectivity can receive the open-source designs for critical elements such as microscopes and print them locally.

3.2 Local industries

3D printing enables endless transformational possibilities, including the creation of new, innovative, locally made unique products which meet specific local needs (Savonen, 2019; Wu et al., 2022). Bespoke products used in cultural events such as local theatres and festivals could be produced from 3D printing since other manufacturing methods might be impractical and/or not economically attractive. 3D printing has been used to digitalise traditional arts and craft processes. For example, needle fleeting is a manual process for making intricate objects such as figurines or putting ornaments to textile objects (Becker, 2022). This process easily lends itself to 3D printing allowing for quicker turnaround time for prototyping of physical objects while also supporting a high level of customisation to be used with different types of materials (Hudson, 2014). 3D printing can be used to make the local practices in low-income communities more efficient. For example, a customised machete peeler was made for a community engaged in peeling tubers in Kenya. Using an ordinary machete to peel the skin of tubers results in removing a significant amount of the produce. The customised machete peeler was 3D printed according to the blade type and thickness typical to the community, with the angle at which the cutting edge is presented customised according to the product being peeled. Another example is the non-electric milk cooler, which is used to mitigate the food storage challenge in rural areas with poor access to electricity. The non-electric cooler maximises the complexities of structures that are possible with 3D printing to create a matrix for the rapid evaporation of water to generate natural cooling. These examples are illustrated in Figure 6.3.

3.3 Medicine

Sub-Saharan Africa is reported to have the worst health care in the world due to the fact that most countries are unable to spend required funds on medical facilities and medicines (IFC, 2022). For over three decades, improvements and innovations in medicine resulting from 3D printing have been documented (Heller et al., 2016). Ishengoma and Mtaho (2014) noted that with access to electricity and

FIGURE 6.3   Examples of 3D-printed products supporting local practices

internet, the adoption of 3D printing can transform the limited access to vital surgical services due to lack of facilities and basic equipment. Several scholars have highlighted the benefits that 3D printing can bring to health care in Africa. Abegaz (2018) noted that 3D technology can bring unprecedented comparative advantages to the health care in Africa. Liaw and Guvendiren (2017) reviewed the applications of 3D printing in medicine and highlighted that current application includes production of medical devices, anatomical models and drug formulation, dentistry and engineered tissue models. They noted that dentistry was one of the advanced fields with application in areas such as restorations, dental models and surgical guides. Another area of wide application of 3D printing is medical devices (both implantable and non-implantable products) such as bone tether plates, hip cups, spinal cages, knee implants, denture bases, craniofacial implants and surgical instruments (Liaw and Guvendiren, 2017).

Examples of functional medical products that have been created using 3D printing include advancement of clinical imaging and reproduction of the human anatomy through structural heart interventions (Vukicevic et al., 2020), production of locally fabricated and low-cost otoscopes to diagnose the prevalence of frequent ear problems (Capobussi and Moja, 2021), the development of a smartphone-based epifluorescence microscope (SeFM) for fresh tissue imaging (Zhu et al., 2020), face shields during pandemic (de Araujo Gomes et al., 2020), medical supplies for children in Haiti (Ishengoma and Mtaho, 2014) and vascularised and perusable cardiac patches production of prosthetic limbs (Abbady et al., 2022; Gretsch et al., 2016; Hofmann et al., 2016).

It should be noted that waste plastics are not suitable for all the applications listed above, especially, implantable devices, due to hygiene and standards; however, non-intrusive components and devices can be made from waste plastics with prosthesis being one of the leading medical applications of 3D printing in the Global South. Furthermore, an advancement in 3D technology will progressively lead to development of appropriate materials as Pugliese et al. (2021) observed that the deficiency of polymers, biomaterials, hydrogels and bioinks was the main drawback in biomedical manufacturing.

3.4 Construction

Construction in Africa is rapidly increasing to meet rising demand and historical deficits (ABP, 2022). There is a growing interest in the application of 3D printing in building and construction. 3D printing has the advantage of creating practical prototypes in rational build time with limited human intervention and the least material wastage (Tay et al., 2017). 3D printing has the potential to revolutionise the construction sector as it can be considered as environment-friendly derived from its limitless possibilities for geometric difficulty in achievability (Hager et al., 2016). El-Sayegh et al. (2020) reviewed 3D printing in construction and noted that the primary advantages of 3D printing in construction include constructability and sustainability benefits, while the drawbacks include material printability, buildability, scalability, structural integrity and lack of codes and regulations.

Hossain et al. (2020) noted that the construction industry is extremely labour-intensive and also a main employment provider and has been undergoing low productivity with minimum technological innovations for decades. 3D printing proffers a possible solution; however, it might not be so welcomed in countries where construction is one of the priority employers and labour is cheaper. However, Buchanan and Gardner (2019) in a review noted that instead of replacing the conventional practice, it can provide a hybrid option with benefits of closer structural efficiency, reduction in material consumption and wastage, streamlining and expedition of the design-build process, enhanced customisation, greater architectural freedom and improved accuracy and safety on-site but new challenges and requirements such as digitally savvy engineers, greater use of advanced computational analysis and a new way of thinking for the design and verification of structures would be required.

4 Conclusion

This chapter highlights the opportunities for waste plastic to be used as a feedstock for 3D printing in sub-Saharan Africa. Given the scale of the plastic challenge, it is fundamentally important to develop innovative solutions to the problem caused by plastic waste. This chapter illustrates that 3D printing coupled with the use of open-source designs can transform low-income societies characterised by underdeveloped infrastructure and inadequate manufacturing capabilities, while addressing the plastic challenge. However, given the infrastructural realities in Africa, it is important to plan around the lack of basic infrastructure such as access to electricity, water and transportation systems. For example, providing off-grid standalone alternate power supply from solar energy. Furthermore, the success of 3D printing at scale in sub-Saharan Africa requires developing capacity and capability of local skills to operate, maintain and develop 3D printing and extruder technology.

Acknowledgement

This work is supported by the United Kingdom Research and Innovation (UKRI) Global Challenges Research Fund (GCRF) under Grant EP/T0238721.

References

Abbady, H.E.M.A., Klinkenberg, E., de Moel, L., Nicolai, N., van der Stelt, M., Verhulst, A.C., Maal, T.J.J., Brouwers, L., 2022. 3D-printed prostheses in developing countries: A systematic review. *Prosthet. Orthot. Int.* 46, 19–30.

Abegaz, S.T., 2018. Marching for 3D printing: Its potential to promoting access to healthcare in Africa, in: Morales-Gonzalez, J.A., Nájera, E.A. (Eds.), *Reflections on Bioethics*. IntechOpen: London, pp. 123–135. https://doi.org/10.5772/intecho pen.75649

ABP, 2022. Construction Activity in Africa Increasing. Africa Bus. https://news.africa-business.com/post/construction-industry-in-africa-building-materials (Accessed 30 November 2022).

Adefila, A., Abuzeinab, A., Whitehead, T., Oyinlola, M., 2020. Bottle house: Utilising appreciative inquiry to develop a user acceptance model. *Built Environ. Proj. Asset Manag.* 10, 567–583. https://doi.org/10.1108/BEPAM-08-2019-0072

Amankwaa, G., Heeks, R., Browne, A.L., 2021. Digital innovations and water services in cities of the global south: A systematic literature review. *Water Altern.* 14, 619–644. www.water-alternatives.org/index.php/alldoc/articles/vol14/v14issue2/637-a14-2-15

Annunziata, M., Bell, G., Buch, R., Patel, S., 2015. *Powering the Future: Leading the Digital Transformation of the Power Industry*. General Electric Company: Boston, MA.

Araujo Galvão, G.D., de Nadae, J., Clemente, D.H., Chinen, G., de Carvalho, M.M., 2018. Circular economy: Overview of barriers. *Procedia CIRP* 73, 79–85. https://doi.org/10.1016/j.procir.2018.04.011

Baechler, C., DeVuono, M., Pearce, J.M., 2013. Distributed recycling of waste polymer into RepRap feedstock. *Rapid Prototyp. J.* 19(2), 118–125.

Barrie, J., Anantharaman, M., Oyinlola, M., Schröder, P., 2022. The circularity divide: What is it? And how do we avoid it? *Resour. Conserv. Recycl.* 180, 106208. https://doi.org/10.1016/J.RESCONREC.2022.106208

Becker, M., 2022. Felt-Concrete Composites in Architecture and Design, in: Open Conference Proceedings, 1, p. 115. https://doi.org/10.52825/ocp.v1i.84

Berg, A., Antikainen, R., Hartikainen, E., Kauppi, S., Kautto, P., Lazarevic, D., Piesik, S., Saikku, L., 2018. Circular Economy for Sustainable Development. Finnish Environment Institute. http://hdl.handle.net/10138/251516

Berman, A., Deuermeyer, E., Nam, B., Chu, S.L., Quek, F., 2018. Exploring the 3D printing process for young children in curriculum-aligned making in the classroom, in: Proceedings of the 17th ACM Conference on Interaction Design and Children, pp. 681–686.

Berman, B., 2012. 3-D printing: The new industrial revolution. *Bus. Horiz.* 55, 155–162.

Buchanan, C., Gardner, L., 2019. Metal 3D printing in construction: A review of methods, research, applications, opportunities and challenges. *Eng. Struct.* 180, 332–348.

Capobussi, M., Moja, L., 2021. An open-access and inexpensive 3D printed otoscope for low-resource settings and health crises. *3D Print. Med.* 7, 1–8.

Celik, E., 2020. *Additive Manufacturing: Science and Technology.* De Gruyter. https://doi.org/10.1515/9781501518782

Chong, L., Ramakrishna, S., Singh, S., 2018. A review of digital manufacturing-based hybrid additive manufacturing processes. *Int. J. Adv. Manuf. Technol.* 95, 2281–2300.

Crowe, C.D., Hendrickson-Stives, A.K., Kuhn, S.L., Jackson, J.B., Keating, C.D., 2021. Designing and 3D printing an improved method of measuring contact angle in the middle school classroom. *J. Chem. Educ.* 98, 1997–2004.

Cruz Sanchez, F.A., Boudaoud, H., Hoppe, S., Camargo, M., Sanchez, F.A.C., Boudaoud, H., Hoppe, S., Camargo, M., 2017. Polymer recycling in an open-source additive manufacturing context: Mechanical issues. *Addit. Manuf.* 17, 87–105. https://doi.org/10.1016/J.ADDMA.2017.05.013

de Araujo Gomes, B., Queiroz, F.L.C., de Oliveira Pereira, P.L., Barbosa, T.V., Tramontana, M.B., Afonso, F.A.C., dos Santos Garcia, E., Borba, A.M., 2020. In-house three-dimensional printing workflow for face shield during COVID-19 pandemic. *J. Craniofac. Surg.* 31(6), e652–e653. doi: 10.1097/SCS.0000000000006723.

Del Rosario, M., Heil, H.S., Mendes, A., Saggiomo, V., Henriques, R., 2022. The field guide to 3D printing in optical microscopy for life sciences. *Adv. Biol.* 6, 2100994.

Devarajan, S., 2010. More cell phones than toilets [WWW Document]. Africa Can End Poverty: A Blog About Econ. Challenges Oppor. Facing Africa. https://blogs.worldbank.org/africacan/more-cell-phones-than-toilets

Duran, A., Parker, J., 2021. How the United Nations International Year of Glass 2022 arrived and what happens now. *Glas. Technol. J. Glas. Sci. Technol. Part A* 62, 45–46.

El-Sayegh, S., Romdhane, L., Manjikian, S., 2020. A critical review of 3D printing in construction: Benefits, challenges, and risks. *Arch. Civ. Mech. Eng.* 20, 1–25.

Feeley, S.R., Wijnen, B., Pearce, J.M., 2014. Evaluation of potential fair trade standards for an ethical 3-D printing filament. *J. Sustain. Dev.* 7(5), 1–12.

Ford, S., Minshall, T., 2019. Invited review article: Where and how 3D printing is used in teaching and education. *Addit. Manuf.* 25, 131–150. https://doi.org/10.1016/J.ADDMA.2018.10.028

Gall, M., Wiener, M., Chagas de Oliveira, C., Lang, R.W., Hansen, E.G., 2020. Building a circular plastics economy with informal waste pickers: Recyclate quality, business model, and societal impacts. *Resour. Conserv. Recycl.* 156, 104685. https://doi.org/10.1016/j.resconrec.2020.104685

Garcia, J., Yang, Z., Mongrain, R., Leask, R.L., Lachapelle, K., 2018. 3D printing materials and their use in medical education: A review of current technology and trends for the future. *BMJ Simul. Technol. Enhanc. Learn.* 4, 27.

Garmulewicz, A., Holweg, M., Veldhuis, H., Yang, A., 2016. Redistributing material supply chains for 3D printing. Proj. Report. Available online www.ifm.eng.cam.ac.uk/uploads/Research/TEG/Redistributing_material_supply_ Chain.pdf (Accessed July 16, 2019).

Gretsch, K.F., Lather, H.D., Peddada, K. V, Deeken, C.R., Wall, L.B., Goldfarb, C.A., 2016. Development of novel 3D-printed robotic prosthetic for transradial amputees. *Prosthet. Orthot. Int.* 40, 400–403. https://doi.org/10.1177/0309364615579317

Hager, I., Golonka, A., Putanowicz, R., 2016. 3D printing of buildings and building components as the future of sustainable construction? *Procedia Eng.* 151, 292–299.

Heller, M., Bauer, H.-K., Goetze, E., Gielisch, M., Roth, K.E., Drees, P., Maier, G.S., Dorweiler, B., Ghazy, A., Neufurth, M., 2016. Applications of patient-specific 3D printing in medicine. *Int. J. Comput. Dent.* 19, 323–339.

Hofmann, M., Harris, J., Hudson, S.E., Mankoff, J., 2016. Helping hands: Requirements for a prototyping methodology for upper-limb prosthetics users, in: Proceedings of the 2016 CHI Conference on Human Factors in Computing Systems, pp. 1769–1780.

Hossain, M.A., Zhumabekova, A., Paul, S.C., Kim, J.R., 2020. A review of 3D printing in construction and its impact on the labor market. *Sustainability* 12, 8492.

Hudson, S.E., 2014. Printing teddy bears: A technique for 3D printing of soft interactive objects, in: Proceedings of the SIGCHI Conference on Human Factors in Computing Systems, pp. 459–468.

IFC, 2022. Health care in Africa: IFC report sees demand for investment [WWW Document]. www.ifc.org/wps/wcm/connect/news_ext_content/ifc_external_cor porate_site/news+and+events/healthafricafeature (Accessed November 30, 2022).

Ishengoma, F.R., Mtaho, A.B., 2014. 3D printing: Developing countries perspectives. *Int. J. Comput. Appl.* 104, 30–34. https://doi.org/10.5120/18249-9329

Kingiri, A.N., Fu, X., 2019. Understanding the diffusion and adoption of digital finance innovation in emerging economies: M-Pesa money mobile transfer service in Kenya. *Innov. Dev.* https://doi.org/10.1080/2157930X.2019.1570695

KLM, 2019. From drink to ink – KLM makes tools from PET bottles [WWW Document]. https://news.klm.com/from-drink-to-ink--klm-makes-tools-from-pet-bottles/ (Accessed November 10, 2021).

Kolade, O., Adegbile, A., Sarpong, D., 2022a. Can university-industry-government collaborations drive a 3D printing revolution in Africa? A triple helix model of technological leapfrogging in additive manufacturing. *Technol. Soc.* 69, 101960. https:// doi.org/10.1016/j.techsoc.2022.101960

Kolade, O., Odumuyiwa, V., Abolfathi, S., Schröder, P., Wakunuma, K., Akanmu, I., Whitehead, T., Tijani, B., Oyinlola, M., 2022b. Technology acceptance and readiness of stakeholders for transitioning to a circular plastic economy in Africa. *Technol. Forecast. Soc. Change* 183, 121954. https://doi.org/10.1016/J.TECHFORE.2022.121954

Kreiger, M.A., Mulder, M.L., Glover, A.G., Pearce, J.M., 2014. Life cycle analysis of distributed recycling of post-consumer high density polyethylene for 3-D printing filament. *J. Clean. Prod.* 70, 90–96. https://doi.org/10.1016/J.JCLEPRO.2014.02.009

Kumar, A., Kumari, K., Sadasivam, R., Goswami, M., 2021. Development of a 3D printer-scanner hybrid from e-waste. *Int. J. Environ. Sci. Technol.* https://doi.org/10.1007/s13 762-021-03131-6

Lawrence, K., 2021. When "Just-In-Time" Falls Short: Examining the Effects of the Suez Canal Blockage, in: *SAGE Business Cases*. SAGE Publications: SAGE Business Cases Originals.

Liaw, C.-Y., Guvendiren, M., 2017. Current and emerging applications of 3D printing in medicine. *Biofabrication* 9, 24102.

Murray, A., Skene, K., Haynes, K., 2017. The circular economy: An interdisciplinary exploration of the concept and application in a global context. *J. Bus. Ethics* 140, 369–380. https://doi.org/10.1007/s10551-015-2693-2

Mwanza, B.G., Telukdarie, A., Mbohwa, C., 2018. Impact of socioeconomic factors on the levers influencing households' participation in recycling programs in Zambia, in: 2018 IEEE International Conference on Industrial Engineering and Engineering Management (IEEM). IEEE, pp. 1021–1025.

O'Connell, J., Haines, J., 2022. How much does a 3D printer cost in 2022? [WWW Document]. All3DP. https://all3dp.com/2/how-much-does-a-3d-printer-cost/ (Accessed July 31, 2022).

Obwoge, M.E., Mainya, N.O., Mosoti, D., 2018. Opportunities and challenges of application of 3D printing technology in teaching and learning in developing countries in Africa. *Int. J. Sci. Res.* 7(1), 1859–1862. DOI 10.21275/ART20179745 2319-7064.

Oke, A., Fernandes, F.A.P., 2020. Innovations in teaching and learning: Exploring the perceptions of the education sector on the 4th industrial revolution (4IR). *J. Open Innov. Technol. Mark. Complex.* 6, 31. https://doi.org/10.3390/joitmc6020031

Owen, J., 2018. 3D printed microscopes for STEM teaching in Kenya [WWW Document]. LinkedIn. www.linkedin.com/pulse/3d-printed-microscopes-stem-teaching-kenya-julia-jule-owen/ (Accessed August 28, 2019).

Oyinlola, M., Kolade, O., Schroder, P., Odumuyiwa, V., Rawn, B., Wakunuma, K., Sharifi, S., Lendelvo, S., Akanmu, I., Mtonga, R., Tijani, B., Whitehead, T., Brighty, G., Abolfathi, S., 2022b. A socio-technical perspective on transitioning to a circular plastic economy in Africa. *SSRN Electron. J.* https://doi.org/10.2139/ssrn.4332904

Oyinlola, M., Okoya, S.A., Whitehead, T., Evans, M., Lowe, A.S., 2023. The potential of converting plastic waste to 3D printed products in Sub-Saharan Africa. *Resour. Conserv. Recycl. Adv.* 17, 200129. https://doi.org/10.1016/j.rcradv.2023.200129

Oyinlola, M., Schröder, P., Whitehead, T., Kolade, S., Wakunuma, K., Sharifi, S., Rawn, B., Odumuyiwa, V., Lendelvo, S., Brighty, G., Tijani, B., Jaiyeola, T., Lindunda, L., Mtonga, R., Abolfathi, S., 2022b. Digital innovations for transitioning to circular plastic value chains in Africa. *Africa J. Manag.* 8, 83–108. https://doi.org/10.1080/23322373.2021.1999750

Paio, A., Eloy, S., Rato, V.M. et al., 2012. Prototyping vitruvius, new challenges: Digital education, research and practice, *Nexus Netw J.* 14, 409–429. https://doi.org/10.1007/s00004-012-0124-6.

Pavlo, S., Fabio, C., Hakim, B., Mauricio, C., 2018. 3D-printing based distributed plastic recycling: A conceptual model for closed-loop supply chain design, in: 2018 IEEE International Conference on Engineering, Technology and Innovation (Ice/Itmc). IEEE, pp. 1–8.

Pinger, C.W., Geiger, M.K., Spence, D.M., 2019. Applications of 3D-printing for improving chemistry education. *J. Chem. Educ.* 97, 112–117.

Pugliese, R., Beltrami, B., Regondi, S., Lunetta, C., 2021. Polymeric biomaterials for 3D printing in medicine: An overview. *Ann. 3D Print. Med.* 2, 100011.

Ragaert, K., Delva, L., Van Geem, K., 2017. Mechanical and chemical recycling of solid plastic waste. *Waste Manag.* 69, 24–58.

Ramalingam, B., Hernandez, K., Prieto Martín, P., Faith, B., 2016. *Ten Frontier Technologies for International Development*. IDS, Brighton.

Rauch, E., Dallasega, P., Matt, D.T., 2016. Sustainable production in emerging markets through Distributed Manufacturing Systems (DMS). *J. Clean. Prod.* 135, 127–138. https://doi.org/10.1016/J.JCLEPRO.2016.06.106

Rogge, M.P., Menke, M.A., Hoyle, W., 2017. 3D printing for low-resource settings. *Bridg.* 47, 37–45.

Ryberg, M.W., Hauschild, M.Z., Wang, F., Averous-Monnery, S., Laurent, A., 2019. Global environmental losses of plastics across their value chains. *Resour. Conserv. Recycl.* 151, 104459.

Sanchez, F.A.C., Boudaoud, H., Camargo, M., Pearce, J.M., 2020. Plastic recycling in additive manufacturing: A systematic literature review and opportunities for the circular economy. *J. Clean. Prod.* 264, 121602.

Santander, P., Cruz Sanchez, F.A., Boudaoud, H., Camargo, M., 2020. Closed loop supply chain network for local and distributed plastic recycling for 3D printing: A MILP-based optimization approach. *Resour. Conserv. Recycl.* 154, 104531. https://doi.org/10.1016/J.RESCONREC.2019.104531

Santos, I.M., Ali, N., Areepattamannil, S., 2019. Interdisciplinary and international perspectives on 3D printing in education, in: Lantz, J. (Ed.), *Girls and 3D Printing: Considering the Content, Context, and Child* (pp. 134–157). IGI Global.

Savonen, B.L., 2019. *A Methodology for Triaging Product Needs for Localized Manufacturing with 3D Printing in Low-Resource Environments.* The Pennsylvania State University.

Schelly, C., Anzalone, G., Wijnen, B., Pearce, J.M., 2015. Open-source 3-D printing technologies for education: Bringing additive manufacturing to the classroom. *J. Vis. Lang. Comput.* 28, 226–237.

Schroeder, P., Oyinlola, M., Barrie, J., Bonmwa, F., Abolfathi, S., 2023. Making policy work for Africa's circular plastics economy. *Resour. Conserv. Recycl.* 190, 106868. https://doi.org/10.1016/j.resconrec.2023.106868

Shah, J., Snider, B., Clarke, T., Kozutsky, S., Lacki, M., Hosseini, A., 2019. Large-scale 3D printers for additive manufacturing: Design considerations and challenges. *Int. J. Adv. Manuf. Technol.* 104, 3679–3693.

Simons, A., Avegnon, K.L.M., Addy, C., 2019. Design and development of a delta 3D printer using salvaged e-waste materials. *J. Eng. (United Kingdom)* 2019, 9. https://doi.org/10.1155/2019/5175323

Singh, N., Hui, D., Singh, R., Ahuja, I.P.S., Feo, L., Fraternali, F., 2017. Recycling of plastic solid waste: A state of art review and future applications. *Compos. Part B Eng.* 115, 409–422.

Swiss Business Hub, 2018. Silicon Savannah: Tapping the potential of Africa's Tech Hub, GLOBAL OPPORTUNITIES. https://www.s-ge.com/en/article/global-opportunities/20213-c6-kenya-tech-hub-fint1 (Accessed 3 May 2021).

Syngenta, 2019. How can digital solutions help to feed a growing world? Available at: www.syngentafoundation.org/file/12811/download (Accessed 13 May 2021).

Tay, Y.W.D., Panda, B., Paul, S.C., Noor Mohamed, N.A., Tan, M.J., Leong, K.F., 2017. 3D printing trends in building and construction industry: A review. *Virtual Phys. Prototyp.* 12, 261–276.

Thompson, R.C., Moore, C.J., Vom Saal, F.S., Swan, S.H., 2009. Plastics, the environment and human health: Current consensus and future trends. *Philos. Trans. R. Soc. B Biol. Sci.* 364, 2153–2166.

Vukicevic, M., Filippini, S., Little, S.H., 2020. Patient-specific modeling for structural heart intervention: Role of 3D printing today and tomorrow CME. *Methodist Debakey Cardiovasc. J.* 16, 130.

Wabnitz, C., Nichols, W.J., 2010. Plastic pollution: An ocean emergency. *Mar. Turt. Newsl.* 1, 1–4.

Woern, A.L., McCaslin, J.R., Pringle, A.M., Pearce, J.M., 2018. RepRapable Recyclebot: Open source 3-D printable extruder for converting plastic to 3-D printing filament. *HardwareX* 4, e00026.

Wu, H., Mehrabi, H., Karagiannidis, P., Naveed, N., 2022. Additive manufacturing of recycled plastics: Strategies towards a more sustainable future. *J. Clean. Prod.* 335, 130236. https://doi.org/10.1016/J.JCLEPRO.2021.130236

Zander, N.E., Gillan, M., Lambeth, R.H., 2018. Recycled polyethylene terephthalate as a new FFF feedstock material. *Addit. Manuf.* 21, 174–182. https://doi.org/10.1016/J.ADDMA.2018.03.007

Zhong, S., Pearce, J.M., 2018. Tightening the loop on the circular economy: Coupled distributedrecyclingandmanufacturingwithrecyclebotandRepRap3-Dprinting. *Resour. Conserv. Recycl.* 128, 48–58. https://doi.org/10.1016/J.RESCONREC.2017.09.023

Zhu, W., Pirovano, G., O'Neal, P.K., Gong, C., Kulkarni, N., Nguyen, C.D., Brand, C., Reiner, T., Kang, D., 2020. Smartphone epifluorescence microscopy for cellular imaging of fresh tissue in low-resource settings. *Biomed. Opt. Express* 11, 89–98.

7

BLOCKCHAINS FOR CIRCULAR PLASTIC VALUE CHAINS

Oluwaseun Kolade

1 Introduction

The plastic value chain comprises four key phases: design, production, use and end of life. An effective discussion of the merits and potentials of the circular plastic economy must be underpinned by a whole value chain approach. Currently, most of the scholarly research on plastic pollution, and much of the ongoing campaigns to tackle the same, have typically focused on the end-of-life phase (Johansen et al., 2022). However, the problems and potentials of plastic waste do not begin at the end-of-life phase, where the efforts are effectively restricted to mitigation rather than prevention and control of the plastic waste problem. This chapter therefore begins with a review of the plastic value chain, with critical reflections on the different and comparative implications, at each of the phases, for the linear and circular economy models.

The plastic value chain begins at the design phase. The design of plastic products, or plastic parts in composite manufactured products, typically follows the Design for Manufacture and Assembly (DFMA) framework. Designers often use the injection moulding process which allows the manufacture of custom products and components of varying sizes and thickness (Karania et al., 2004). It is during the design phase that key decisions are taken about the operational features, functional properties and quality of the final product. These include questions of polymer mix and recyclability, which are key considerations for circularity of the final product. The traditional paradigm for design of plastic products allows designers to mix different polymers, as well as incorporate various combinations of additives, coolants and adhesives in the manufacture of plastic products. The resulting final products are therefore too customised and complex that they are hardly suitable for recycling (Plastic Ocean, 2022). The design of circular plastic products should therefore be based on standardised, simpler materials and fewer

DOI: 10.4324/9781003278443-9

polymer types in order to enhance their future recyclability (Johansen et al., 2022). Thus, ecodesign of circular plastic products are undergirded by five key principles: design for sustainable sourcing, design for optimised resource use, design for environmentally sound and safe product use, design for prolonged product use and design for recycling (Foschi et al., 2020).

The production phase of the plastic value chain includes the extraction and production of raw materials, the production of primary plastics and the production of secondary plastics. The vast majority of plastic products – with some estimates putting it at 99% – are currently derived from petrochemicals sourced from fossil fuels, with the remaining estimated 1% produced from bio-based materials (James, 2017). As national policies and multilateral initiatives gather momentum to restrict the use of petrochemicals, the proportion of renewable biomaterials used in making plastics is likely to increase in the future. Primary plastic production focuses on the production of primary plastic pellets (primary microplastics), which are either produced as monomers for in-house conversion by large companies or otherwise sold to other companies to polymerise at a smaller scale (James, 2017; Ryberg et al., 2018). The production of plastic products is undertaken through the melting and moulding of primary plastics (Johansen et al., 2022). The five types of primary plastics that account for most of the global plastic production are polyethylene terephthalate (PET), high-density polyethylene (HDPE), polyvinyl chloride (PVC), low-density polyethylene (LDPE) and polypropylene (PP) (James, 2017). The production of secondary plastics relates essentially to plastic waste management via recycling of primary plastics.

The use phase of the plastic value chain spans a wide spectrum of sectors including containers and packaging, engineering and construction, consumer goods, industrial machinery, transportation and textiles (James, 2017). The activities in this phase include demand and purchase, use and post-consumption handling of plastic products (Johansen et al., 2022). Demand and purchase patterns are often influenced by consumers' levels of awareness and understanding of environmental impact of the plastic products and knowledge of alternative, more environment-friendly products (Boesen et al., 2019). Policy and regulatory factors also influence societal habits of consumption, for example, through incentives and support for alternative products and packaging (Kolade et al., 2022b; Oyinlola et al., 2022) and through various levies and taxes on plastic bags, especially single-use plastics (Syberg et al., 2021).

The end-of-life phase is the phase that has currently attracted the most attention in the plastic value chain. This phase includes a wide range of activities, processes and factors, including collection and sorting, recycling, life cycle assessment and policy and regulations (Johansen et al., 2022). Increasing the rate of plastic waste collection is an important step needed to divert plastic wastes from landfills, thereby ensuring cleaner waste streams (Syberg et al., 2021). In order to realise this desirable outcome for a circular plastic economy, the responsibility for collection needs to be shared by producers and incentivised by appropriate regulations (European Union, 2018). One of the instruments that has been used to encourage collection and subsequent recycling of plastic products is the deposit refund system (DRS). This is a market-based instrument in which the consumer is required to pay an extra amount of money as deposit for product packaging at the point of purchase, and this deposit is then refunded at the point of return of the container (Sanabria Garcia and Raes, 2021). An effective and efficient plastic waste collection system is an important first step that determines the success of subsequent activities, including sorting and recycling.

Plastic recycling can be either mechanical or chemical. The main difference between the physical and chemical methods of plastics recycling is that physical recycling does not entail any alteration of the structure and composition of the polymer material of which the plastic product is composed (Martinez Sanz et al., 2022). On the other hand, chemical recycling typically involves changes in the polymer structure, including depolymerisation of polymers into monomers, from which they are subsequently purified and returned into the polymerisation process towards the making of new products. As mentioned above, the original structure of the initial product comes into play at this point, as complex polymers, including those with additives and adhesives, are much more difficult to chemically break down in the process of making new products. The more viable option for such products is physical recycling, although this in effect offers more limited product options in recycling.

The preceding review underlines the intricate linkages across the different phases of the plastic value chain (see Figure 7.1 for an overview of the four phases). The design of plastic products around simpler, more standardised polymer structures has a direct impact on the recyclability of the final product. In terms of production, the drive towards the use of renewable biomaterials in the production of biodegradable plastics can make an increasingly substantial contribution to the global campaign to reduce plastic wastes (Goel et al., 2021). The specific structure of the material produced also influences the type of recycling methods suitable for the end-of-life phase of the product. Thermoplastic polymers, for example, can be easily recycled, either through mechanical processes or chemical processes such as chemolysis, cracking or gasification (Morici et al., 2022). On the other hand, thermoset plastics are more difficult to recycle because they are heat and chemical resistant and usually require high energy input.

Given the foregoing, a multi-stakeholder, multi-sectoral approach is required, across the key phases of the plastic value chain, to accelerate the transition to a circular plastic economy. Digital innovations can play a key role in facilitating this collaborative synergy of stakeholders. This chapter focuses attention on blockchains as an especially auspicious Industry 4.0 technology that potentially has applicability across the entire plastic value chain. In order to explicate this, this chapter takes a conceptual approach with a case illustration to explore the merits and limits of blockchains as a driver of a circular plastic economy on the African continent. The rest of this chapter is organised as follows: first, this chapter provides a review of the relevance and application of blockchains in the circular economy, before zeroing in on the applications of blockchains in the plastic value chain. This

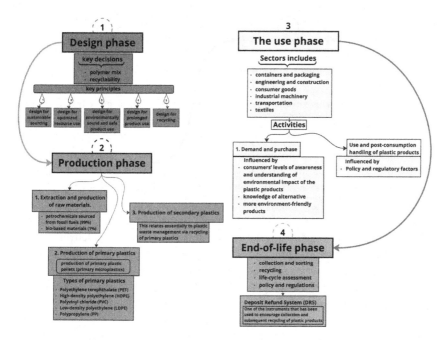

FIGURE 7.1   The four phases of the plastic value chain (author)

is then followed with a case study of BanQu, a blockchain solution launched in partnership with Coca-Cola Africa to improve local recycling and drive a circular plastic economy in South Africa. This chapter ends with a discussion, conclusion and recommendations.

2   Blockchains, circular economy and plastic value chains

2.1   *Blockchains and the circular economy*

Blockchains are defined as "tamper evident and tamper resistant digital ledgers implemented in a distributed fashion and usually without a central authority" (Yaga, Mell, Roby, and Scarfone, 2018, p. iv). Blockchains are thus characterised by the principles of decentralisation, persistency, anonymity and auditability (Zheng et al., 2017). In place of a centralised third-party intermediaries (such as central banks), blockchains deploy consensus algorithms to maintain data consistency in a distributed system (Kolade et al., 2022a). The anonymity, autonomy and interoperability of blockchains enable separate parties to efficiently and seamlessly share data and synchronise their services in a process that is tamper resistant but does not require trust among the parties (Sanka et al., 2021).

Although it was originally developed within the context of financial systems (Lee, 2019), blockchains have gained increasing traction among scholars, practitioners and campaigners for the circular economy. It has been noted that a digital solution that enables transparent sharing of information about materials and supply chains can facilitate more circular resource flows (Böckel et al., 2021). It can also facilitate a meeting point of stakeholders and a melting pot of ideas to accelerate the transition to a circular economy.

With regard to the supply chains, blockchain technology can enable the transformation of traditional supply chains to circular supply chains in order to optimise resource allocation and promote sustainability (Huang et al., 2022). Blockchains integrate the three key supply chain reverse processes – recycle, redistribute and re-manufacture – with the three key factors that underpin blockchain technology: trust, traceability and transparency (Centobelli et al., 2022). It can be used to manage, share and monitor key product information such as quality, quantity, location and ownership (Centobelli et al., 2022) and other important parameters such as product demand, transaction price, delivery period, resource recycling rate and greenhouse gas emissions (Huang et al., 2022). These are all of critical interest to supply chain stakeholders.

In addition to supply chain applications, blockchains have also been deployed to create and manage new circular economy ecosystems through the use of tokens (Narayan and Tidström, 2020). Tokens are digital assets that can be transferred between parties in a decentralised system, and they can be exchanged for fiat currencies. Blockchain tokens can be used to incentivise consumers to return products or recycle wastes (Rejeb et al., 2022). Tokens can therefore be used to integrate otherwise disconnected product ecosystems and thereby support the transition from linear, competitive models of value creation and appropriation to circular, co-opetitive models of value circulation among stakeholders (Narayan and Tidström, 2020). Incentivisation opportunities, created via tokens, are used to support new product uptake, testing and validation (Nandi et al., 2021). They can also be used to integrate actors from low-income communities and poorer households to the circular economy.

2.2 Applications of blockchains in the plastic value chain

Following on from the discussion of the general merits of blockchains in the circular economy, this section now turns attention to the application of blockchain in the plastic value chain. The introductory section has identified the four key phases of the plastic value chain, namely the design phase, the production phase, the use phase and the end-of-life phase. This section will highlight and discuss the applicability and potentials of blockchains in each of these four phases (Figure 7.2).

2.2.1 The design phase

One of the main talking points in the drive towards a circular plastic economy is the challenge of holding plastic manufacturers to account on promises to

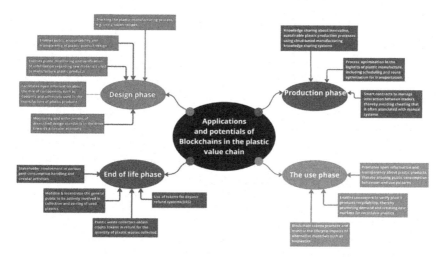

FIGURE 7.2   Blockchain applications in the plastic value chain (author)

support circularity and sustainability. This challenge is directly related to the lack of transparency about material composition, starting with the design of plastic products. Blockchain tokens can be used to track the plastic manufacturing process, including the composition of raw materials that underpin the design of plastic products. One innovative idea that has been recently proposed is the use of token recipes. This concept is based on representing physical materials, including raw materials, as digital tokens, and then identifying recipes that are used to transform the physical products (Westerkamp et al., 2020). As well as capturing the stages of tokenisation and recipes for product transformation, the process also includes certification of products to verify product standardisation and monitoring of compliance with regulations and quality control. For the design phase of the plastic value chain, token recipes, enabled by blockchain technology, enable public accountability and transparency of plastic product design. Other blockchain-enabled solutions that have been proposed include molecular tagging and digital product passport, through which plastic products can be traced from start to finish in a system that is immutable, transparent, secure and efficient (Bhubalan et al., 2022). In combination, these blockchain solutions enable public monitoring and verification of information regarding raw materials used to manufacture plastic products and open information about the mix of components such as coolants and adhesives that have been used in the manufacture of plastic products. It also enables regulators to monitor and enforce prescribed design standards in the drive towards a circular economy. Finally, transparent product design help to mitigate, if not entirely eliminate, one of the main challenges faced by other stakeholders further the value chain. This is achieved through verifiable product information that help in reuse, recycling and redesign of plastic products.

2.2.2   The production phase

The production phase of the plastic value chain is closely intertwined with the design phase. It entails physical and chemical combinations of polymers, along with other additives, adhesives and coolants, to manufacture products. In addition to information transparency about these constituent components, blockchain technology provides new opportunities for knowledge owners to share intellectual assets about novel, sustainable plastic production processes using, for example, cloud-based manufacturing knowledge-sharing systems (Li et al., 2018). In addition, blockchains can enable process optimisation in the logistics of plastic manufacturing process, including optimisation of transportation route maps and schedules, and the use of smart contracts to manage interaction between traders, thereby avoiding cheating that is often associated with manual systems (Xu and He, 2022).

2.2.3   The use phase

While entrenched societal habits of plastic use is influenced by the ubiquity of plastic products in the linear economy, consumption behaviour is also related to public awareness of, and ease of access to information about, the severe environmental impacts of plastic wastes. Thus, blockchain, by promoting open information and transparency about plastic products, can shape public consumption behaviour and use patterns (Boesen et al., 2019). One example is the Plastic Credit, a blockchain solution that enables consumers to verify the recyclability of plastic products, thereby promoting demand and creating new markets for recyclable plastics (Liu et al., 2021). Blockchain tokens can also promote and monitor the life cycle impacts of alternative materials such as bioplastics (Gerassimidou et al., 2022).

2.2.4   End-of-life phase

As mentioned in the introduction, the end-of-life phase is the phase in the plastic value chain that has attracted the biggest attention from a wide range of stakeholders. It is therefore the phase where blockchain technologies are finding the most active applications. Blockchain solutions are being used to involve stakeholders in various post-consumption handling and circular activities, including plastic waste collection, sorting, reuse, recycling and re-manufacture, among others. Blockchain technologies are being proposed to organise deposit refund systems (Reloop Platform, 2022). Blockchain tokens are also being used to mobilise and incentivise the general public, especially in low-income communities to be actively involved in collection and sorting of used plastics (Verma et al., 2022). Either in their own households or across the community, plastic waste collectors obtain crypto tokens in return for the quantity of plastic wastes collected. These tokens can then be exchanged for cash or used to access services.

3   Case study of BanQu, South Africa

3.1   Background

BanQu was founded in 2015 by Ashish Gadnis, a serial tech entrepreneur who was at the time working as a volunteer for the US Agency for International Development in the Democratic Republic of Congo. There, he had met a mother who could not secure a loan to pay her children's tuition. She was barely a farmer with considerable assets, but the bank rejected her scratch paper receipts as proof of sale for her loan application (Zhong, 2019). Gadnis recognised there the potential of digital currency to address the fundamental flaw in the global financial system that has, in effect, excluded a significant population of informal and micro entrepreneurs in developing countries. His brainchild, BanQu, is a blockchain-enabled supply chain solution which offers microenterprises and large organisations the opportunity to track and trace end-to-end transactions, covering "every mile", in a secure and transparent system (BanQu, 2022a). Among others, this digital solution enables enterprises to reduce costs and fix issues more quickly with real-time visibility in the supply chains, track raw materials and finished items from source to shelf to salvage and replace manual processes and paper-based documentation with tamper-free digital documentation and processes. BanQu has established itself as a blockchain platform of choice for refugees and people in extreme poverty. It is an accessible digital innovation that is driving integration of informal micro entrepreneurs from poorer and developing countries to the global economy (Sustainable Brands, 2016).

3.2   BanQu application for plastic waste management in Africa

In March 2021, Coca-Cola South Africa in partnership with BanQu launched a "payment platform to financially empower informal waste reclaimers and buyback centres in a boost to the local recycling sector" (Bulbulia, 2021). This initiative recognises the enormous contributions of otherwise invisible informal waste collectors to the circular economy. For more than 60,000 waste reclaimers in South Africa, BanQu solution enables them to create a permanent digital record of transactions, thereby enabling them to demonstrate their earnings in order to access credit. The BanQu solution enables low-risk, cashless transactions and generates public awareness about the contributions of waste collectors and buyback centres.

The BanQu payment solution is low tech in terms of accessibility, as users do not require expensive smartphones, and transactions are communicated via simple SMS. Users have given testimonials about the functional benefits they derive from the platform, including data-driven, strategic management of business operations. One recycling business owner notes:

> I thought the app would be a lot of work, but it makes things much more efficient with regard to data capturing, and the ins and outs and operations of my business, which I think is more important than anything … It allows us to have longevity

because we know how to read the data. What are our trading volumes and how do we maintain them? If we drop, what can we do better? We can go back through our historical data records online and see what worked. I gained business insights that I didn't think I needed but which have come in very useful.

Refilwe Ramadikela, chief executive of Hendrina Recycling
in Mpumalanga, South Africa, July 2022, IT Web Interview;
IT Web, 2022

The platform is also fully integrated with mobile money applications so that reclaimers can store their earnings in secure e-wallets and withdraw cash from ATMs (BanQu, 2022b). BanQu promotes financial inclusion of waste reclaimers and enables them to create financial records and credit history, giving them economic visibility in the global supply chain. It also enables owners of buyback centres to better understand and develop their businesses through the use of automated recording and tracking of transactions (Bulbulia, 2021). The payment solution is gaining traction in South Africa, and one key stakeholder has suggested that the data insights from the growing number of users registered on the platform, and the associated transactions, offer promising prospects for the future of the recyclable waste economy:

With over R10 million worth of transactions representing over 4,000 tonnes of recyclables, we are beginning to reach a point at which our data has sufficient scale and diversity to provide a credible basis for analysis. By looking at this data, we can not only begin to understand the market better but also use these insights to support and grow the informal waste economy.

David Drew, PETCO vice-chairman and Coca-Cola's sustainability
director for Africa July 2022, IT Web Interview (IT Web, 2022)

4   Discussion

The BanQu case study underlines the point made previously that most of the attention and activities in the circular plastic economy has focused on the end-of-life phase of the plastic value chain. The BanQu blockchain solution has provided waste collectors with economic identity and visibility in the ecosystem. This has wide ranging implications, including the enactment of "dignity through identity" for otherwise unrecognised and excluded, yet important, economic actors (BanQu, 2017). The digital innovation has effectively demonstrated that it is possible to obtain better financial rewards from plastic waste collection activities, which many collectors have originally been pushed into because of unemployment, poverty and limited economic opportunities. With this development, stakeholders such as recycling centre owners and policymakers can mobilise more waste collectors and other participants into the circular plastic economy. In Spain, campaigners were able to mobilise citizens using virtual tokens and gamification for greater and more effective participation in the circular economy (Gibovic and Bikfalvi, 2021).

The BanQu technology can also enable other sectors of the wider circular economy through a complementary and mutually reinforcing mechanism. For example, with minimal tweaking, the BanQu solution can be deployed to link households discarding recoverable and reusable non-plastic items with collectors and enterprises who are able to reclaim value from the discarded items through recycling and refurbishment. Examples of such items include furniture, electronics, kitchen and other items that are sometimes discarded by higher income families but can be reclaimed and/or refurbished for use by others.

However, given the potential applications and reach of the BanQu digital solution for traceability and transparency, the technology has been significantly underutilised from the whole-value chain perspective in the South African case described above. There is no evidence that the solution has been applied yet in the earlier phases of the value chain, especially design and production of plastic products. Yet, the technology is highly promising in this area. There are two main factors that could possibly explain the relative lack of progress in these phases. The first is that, as reported, the project is jointly funded and led by Coca-Cola, a major multinational corporation producing plastic and other products. Coca-Cola would arguably be less invested in any application of the technology that will, in effect, demand more transparency and accountability from them regarding the design and production of plastic products. Conversely, investment in applications that incentivise other actors in the ecosystem to clean up the plastic "mess" is likely to be a more attractive area of investment for Coca-Cola. The other factor is the deficit of political will from governments and institutional actors to convene other stakeholders and apply appropriate policy instruments to drive interest in, and applications of, digital solutions for the earlier phases of the plastic value chain (Dokter et al., 2021). In other words, it will likely require the combined efforts of institutional actors and digital innovators, rather than big corporate sponsors, to lead the process through which blockchain solutions are applied and implemented in the earlier phases of the plastic value chain.

5 Conclusion

This chapter highlights the imperative of a whole-value chain approach to transitioning to a circular plastic economy, with blockchain technology as a key driver. Most of the current activities on the circular plastic economy have focused on the latter stages of the plastic value chain, especially the end-of-life phase. In the South African case discussed, blockchain technology has enabled stakeholders to empower otherwise invisible and unrecognised informal waste collectors, helping them to gain economy identity and create credit history, among others. This is significant and pathbreaking in many ways, not least in terms of the potentials to expand the circular plastic ecosystem and integrate millions of informal workers into national economies.

However, in other ways, the achievements on waste collection and recycling could be viewed as prioritising effects over causes or otherwise distracting from more critical causal problems. This is considering the fact that the increased volume of plastic wastes is a direct consequence of unsustainable design and manufacture of plastic products. Against this backdrop, therefore, this chapter argues that blockchain technology is a digital innovation of choice to invigorate the circular plastic campaign and redirect efforts in order to tackle the main causes of plastic pollution at their roots. First, with its key features of traceability and transparency, blockchains provide a platform to track and trace the design and production activities of big plastic producers. This will help consumers and the wider public to assess their level of commitment to the circularity agenda, as well as monitor their compliance with regulations and laws promoting a circular plastic economy. Finally, armed with this information about the design values and manufacturing activities of multinational plastic producers, informed and conscious consumers can take more targeted and concrete actions to enable and encourage manufacturers with high ratings on circularity and sustainable production practices.

References

BanQu, 2017. Dignity through identity [WWW Document]. www.id4africa.com/201 7_event/Presentations/3-5-4_Banqu_Inc._Ashish_Gadnis.pdf (Accessed November 23, 2022).

BanQu, 2022a. BanQu: What we do [WWW Document]. What we do. https://banqu.co/ our-platform-solution-technology/

BanQu, 2022b. Optimizing the circular economy: Coca-Cola and BanQu [WWW Document]. Use cases. https://banqu.co/use-cases/optimizing-the-circular-economy-coca-cola-and-banqu/#:~:text=BanQu's online automated supply chain, boosts their visibility and bankability

Bhubalan, K., Tamothran, A.M., Kee, S.H., Foong, S.Y., Lam, S.S., Ganeson, K., Vigneswari, S., Amirul, A.A., Ramakrishna, S., 2022. Leveraging blockchain concepts as watermarkers of plastics for sustainable waste management in progressing circular economy. *Environ Res* 213, 113631. https://doi.org/10.1016/j.envres.2022.113631

Böckel, A., Nuzum, A.K., Weissbrod, I., 2021. Blockchain for the circular economy: Analysis of the research-practice gap. *Sustain Prod Consum* 25, 525–539. https://doi.org/10.1016/ j.spc.2020.12.006

Boesen, S., Bey, N., Niero, M., 2019. Environmental sustainability of liquid food packaging: Is there a gap between Danish consumers' perception and learnings from life cycle assessment? *J Clean Prod* 210, 1193–1206. https://doi.org/10.1016/j.jcle pro.2018.11.055

Bulbulia, T., 2021. Coca-Cola, BanQu launch payment platform for informal waste reclaimers. Creamer Media's Engineering News.

Centobelli, P., Cerchione, R., Vecchio, P. Del, Oropallo, E., Secundo, G., 2022. Blockchain technology for bridging trust, traceability and transparency in circular supply chain. *Inf Manage* 59, 103508. https://doi.org/10.1016/j.im.2021.103508

Dokter, G., Thuvander, L., Rahe, U., 2021. How circular is current design practice? Investigating perspectives across industrial design and architecture in the transition

towards a circular economy. *Sustain Prod Consum* 26, 692–708. https://doi.org/10.1016/
j.spc.2020.12.032

European Union, 2018. "DIRECTIVE (EU) 2018/851 OF THE EUROPEAN
PARLIAMENT AND OF THE COUNCIL of 30 May 2018 amending Directive
2008/98/EC on waste," pp. 109–140. Available at: https://eur-lex.europa.eu/legal-
content/EN/TXT/?uri=uriserv:OJ.L_.2018.150.01.0109.01.ENG

Foschi, E., Zanni, S., Bonoli, A., 2020. Combining eco-design and LCA as decision-
making process to prevent plastics in packaging application. *Sustainability (Switzerland)*
12, 1–13. https://doi.org/10.3390/su12229738

Gerassimidou, S., Lovat, E., Ebner, N., You, W., Giakoumis, T., Martin, O. V., Iacovidou,
E., 2022. Unpacking the complexity of the UK plastic packaging value chain: A
stakeholder perspective. *Sustain Prod Consum* 30, 657–673. https://doi.org/10.1016/
j.spc.2021.11.005

Gibovic, D., Bikfalvi, A., 2021. Incentives for plastic recycling: How to engage citizens
in active collection. Empirical evidence from Spain. *Recycling* 6, 29. https://doi.org/
10.3390/recycling6020029

Goel, V., Luthra, P., Kapur, G.S., Ramakumar, S.S.V., 2021. Biodegradable/
bio-plastics: Myths and realities. *J Polym Environ.* https://doi.org/10.1007/s10
924-021-02099-1

Huang, L., Zhen, L., Wang, J., Zhang, X., 2022. Blockchain implementation for circular
supply chain management: Evaluating critical success factors. *Ind Mark Manage* 102,
451–464. https://doi.org/10.1016/j.indmarman.2022.02.009

IT Web, 2022. Blockchain pilot project boosts SA's recycling efforts [WWW Document].
Open Source. www.itweb.co.za/content/kYbe9MXb6JjvAWpG

James, G., 2017. The plastics landscape: Risks and opportunities along the value chain.
www.unpri.org/download?ac=10258

Johansen, M.R., Christensen, T.B., Ramos, T.M., Syberg, K., 2022. A review of the
plastic value chain from a circular economy perspective. *J Environ Manage* 302, 113975.
https://doi.org/10.1016/j.jenvman.2021.113975

Karania, R., Kazmer, D., Roser, C., 2004. Plastics product and process design strategies.
Proceedings of the ASME Design Engineering Technical Conference 3, 703–712.
https://doi.org/10.1115/detc2004-57755

Kolade, O., Adepoju, D., Adegbile, A., 2022a. Blockchains and the disruption of
the sharing economy value chains. *Strategic Change* 31, 137–145. https://doi.org/
10.1002/JSC.2483

Kolade, O., Odumuyiwa, V., Abolfathi, S., Schröder, P., Wakunuma, K., Akanmu, I.,
Whitehead, T., Tijani, B., Oyinlola, M., 2022b. Technology acceptance and readiness
of stakeholders for transitioning to a circular plastic economy in Africa. *Technol Forecast
Soc Change* 183. https://doi.org/10.1016/j.techfore.2022.121954

Lee, J.Y., 2019. A decentralized token economy: How blockchain and cryptocurrency
can revolutionize business. *Bus Horiz* 62, 773–784. https://doi.org/10.1016/j.bus
hor.2019.08.003

Li, Z., Liu, L., Barenji, A.V., Wang, W., 2018. Cloud-based manufacturing
blockchain: Secure knowledge sharing for injection mould redesign. *Procedia CIRP* 72,
961–966. https://doi.org/10.1016/j.procir.2018.03.004

Liu, C., Zhang, X., Medda, F., 2021. Plastic credit: A consortium blockchain-based
plastic recyclability system. *Waste Manage* 121, 42–51. https://doi.org/10.1016/j.was
man.2020.11.045

Martinez Sanz, V., Morales Serrano, A., Schlummer, M., 2022. A mini-review of the physical recycling methods for plastic parts in end-of-life vehicles. *Waste Manage Res: J Sustainable Circ Econ* 40, 1757–1765. https://doi.org/10.1177/0734242X221094917

Morici, E., Carroccio, S.C., Bruno, E., Scarfato, P., Filippone, G., Dintcheva, N.T., 2022. Recycled (bio)plastics and (bio)plastic composites: A trade Opportunity in a green future. *Polymers (Basel).* https://doi.org/10.3390/polym14102038

Nandi, S., Sarkis, J., Hervani, A.A., Helms, M.M., 2021. Redesigning supply chains using blockchain-enabled circular economy and COVID-19 experiences. *Sustain Prod Consum* 27, 10–22. https://doi.org/10.1016/j.spc.2020.10.019

Narayan, R., Tidström, A., 2020. Tokenizing coopetition in a blockchain for a transition to circular economy. *J Clean Prod* 263, 121437. https://doi.org/10.1016/j.jclepro.2020.121437

Oyinlola, M., Schröder, P., Whitehead, T., Kolade, O., Wakunuma, K., Sharifi, S., Rawn, B., Odumuyiwa, V., Lendelvo, S., Brighty, G., Tijani, B., Jaiyeola, T., Lindunda, L., Mtonga, R., Abolfathi, S., 2022. Digital innovations for transitioning to circular plastic value chains in Africa. *Africa J. Manage* 8, 83–108. https://doi.org/10.1080/23322373.2021.1999750

Plastic Ocean, 2022. Plastics initiative – the ocean foundation [WWW Document]. Plastic Ocean. https://oceanfdn.org/initiatives/plastics-initiative/#resources (Accessed November 11, 2022).

Rejeb, A., Binti Dato Mohamad Zailani, S.H., Rejeb, K., Treiblmaier, H., Keogh, J.G., 2022. Modeling enablers for blockchain adoption in the circular economy. *Sustainable Futures* 4, 100095. https://doi.org/10.2139/ssrn.4118439

Reloop Platform, 2022. Digital Deposit Return Systems – What you need to know. www.reloopplatform.org/digital-deposit-return-systems-what-you-need-to-know/ (Accessed October 20, 2022)

Ryberg, M.W., Laurent, A., Hauschild, M., 2018. Mapping of global plastics value chain and plastics losses to the environment (with a particular focus on marine environment). United Nations Environment Programme 1–99. https://wedocs.unep.org/20.500.11822/26745

Sanabria Garcia, E., Raes, L., 2021. Economic Assessment of a Deposit Refund System (DRS), an Instrument for the Implementation of a Plastics Circular Economy in Menorca, Spain.

Sanka, A.I., Irfan, M., Huang, I., Cheung, R.C.C., 2021. A survey of breakthrough in blockchain technology: Adoptions, applications, challenges and future research. *Comput Commun.* https://doi.org/10.1016/j.comcom.2020.12.028

Sustainable Brands, 2016. Ashish Gadnis: Founder, BanQu [WWW Document]. SB Insights. https://sustainablebrands.com/is/ashish-gadnis#:~:text=Hewent on to build,soon after%2C BanQu was born

Syberg, K., Nielsen, M.B., Westergaard Clausen, L.P., van Calster, G., van Wezel, A., Rochman, C., Koelmans, A.A., Cronin, R., Pahl, S., Hansen, S.F., 2021. Regulation of plastic from a circular economy perspective. *Curr Opin Green Sustain Chem* 29, 100462. https://doi.org/10.1016/j.cogsc.2021.100462

Verma, D., Okhawilai, M., Dalapati, G.K., Ramakrishna, S., Sharma, A., Sonar, P., Krishnamurthy, S., Biring, S., Sharma, M., 2022. Blockchain technology and AI-facilitated polymers recycling: Utilization, realities, and sustainability. *Polym Compos* 1–15. https://doi.org/10.1002/pc.27054

Westerkamp, M., Victor, F., Küpper, A., 2020. Tracing manufacturing processes using blockchain-based token compositions. *Digital Commun. Networks* 6, 167–176. https://doi.org/10.1016/j.dcan.2019.01.007

Xu, X., He, Y., 2022. Blockchain application in modern logistics information sharing: A review and case study analysis. *Prod Plann Control* 0, 1–15. https://doi.org/10.1080/09537287.2022.2058997

Yaga, D., Mell, P., Roby, N. and Scarfone, K. (2018) Blockchain technology overview. Gaithersburg, MD. Available at: https://doi.org/10.6028/NIST.IR.8202.

Zheng, Z., Xie, S., Dai, H., Chen, X., Wang, H., 2017. An overview of blockchain technology: Architecture, consensus, and future trends, in: 2017 IEEE International Congress on Big Data (BigData Congress). IEEE Computer Society, pp. 557–564. https://doi.org/10.1109/BigDataCongress.2017.85

Zhong, C., 2019. Innovator BanQu builds blockchain and bridges for traceability, small farmers' livelihoods [WWW Document]. *Climate Tech Weekly.* www.greenbiz.com/article/innovator-banqu-builds-blockchain-and-bridges-traceability-small-farmers-livelihoods#:~:text=Gadnis founded BanQu in 2015,up for the free service

8

TRANSITIONING TO A CIRCULAR PLASTIC ECONOMY IN WEST AFRICA THROUGH DIGITAL INNOVATION

Challenges and the Way Forward

Victor Odumuyiwa and Ifeoluwa Akanmu

1 Introduction

Africa has been described as the world's youngest continent because it is the continent that has the highest proportion of its population being youth. A whopping 77% of Africans are aged 35 and below, with the majority of those in this group being between 15 and 25 years. With a median age of 19.7 years, not only does Africa have the youngest population, it also has the highest population growth rate, and by 2055, about 60% of worldwide population growth will come from Africa. Current trends estimate that it will remain the fastest growing continent in the world throughout the remainder of the 21st century, with the youth population more than doubling from current levels by 2055 (UN DESA, 2015; Kariba, 2020). This of course means that consumption levels will soar through the roof, as high population and population growth rates, in combination with other factors such as technological advancements and urbanisation, will increase the general demand for resources. In recent times, household consumption in Africa has risen at a higher rate than gross domestic product (GDP), and its consumer markets are currently one of the world's fastest growing consumer markets (Signé, 2019).

In West Africa, the most urbanised region of sub-Saharan Africa, more and more consumers look for higher value products that are convenient to buy and use. There is an increased use of plastic packaging, and shelf lives of products continue to shrink. All these have the obvious effect of a corresponding increase in the amount of solid waste generated in the region (Staatz & Hollinger, 2016; Torres & Seters, 2016). This is in keeping with trends around the world, as in 2016, according to the World Bank, urban areas generated 2.01 billion tonnes of solid waste, and annual waste generation is anticipated to reach 3.40 billion tonnes in 2050 (Kaza et al., 2018). It has become painfully obvious that the rapid urbanisation and population

DOI: 10.4324/9781003278443-10

increase, as well as the prevalent linear take-make-waste model, have brought in their wake the problem of waste generation and management.

The environmental problem created by uncontrolled plastic waste generation and inefficient waste management in West Africa and Africa at large could lead to great health concerns and disasters in the region and the continent if not adequately and promptly addressed. This chapter seeks to review the progress made so far in West Africa in transitioning to a circular plastic economy (CPE) through digital innovations, identify the challenges inhibiting the transition and also provide some recommendations on the way forward for the region.

2   Plastic Pollution in West Africa

As succinctly put by Bloomberg City Lab (Berg, 2012), "With Urbanization Comes Mountains of Trash". Indeed, one of the most pressing concerns of urbanisation in West Africa has been the problem of solid waste management. An inspection of many African cities today will reveal cluttered roadsides, clogged streams, flooded motorways and accrescent dumpsites. In fact, according to the Waste Atlas 2014 report on world's 50 biggest dumpsites published by D-Waste, 10 of the largest dumpsites in the world are located in West Africa. The number of dumpsites in Nigeria alone is greater than those on the entire European continent (Mavropoulos et al., 2014; Wilson et al., 2015).

This should be surprising, as West Africa belongs to a region that generates the second least amount of waste in the world (Kaza et al., 2018) as shown in Figure 8.1. Studies and reports such as Okafor-Yarwood and Adewumi (2020) and

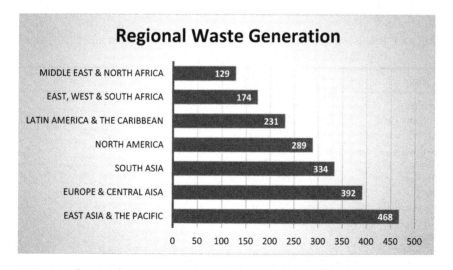

FIGURE 8.1   Regional waste generation in 2016 according to World Bank

Merem et al. (2021) suggest that a lot of the waste in West Africa actually comes from outside the continent, as a result of illegal dumping.

Despite these flagitious activities that definitely contribute significantly towards the amount of waste in the region, it cannot be overlooked that a significant portion of the waste in West Africa is actually generated in West Africa, and the region is well on its way towards an "explosion" in waste generation.

2.1 Plastic and Plastic Waste in West Africa

When it comes to waste, plastic is still at the top of the agenda for many reasons. It is true that most of Africa's land and waterways are increasingly becoming heavily polluted with plastic waste, but beyond being an eyesore or annoyance, plastic waste has been identified as a major contributor towards several problems such as biodiversity loss, greenhouse gas emissions and human health problems. Not only does it constitute an environmental concern in West African countries but it is also a serious socioeconomic issue affecting infrastructure, tourism and livelihoods.

Based on data for 12 West African countries on plastic, the West African region has imported about 28 million tonnes of plastic between 1994 and 2017 (see Table 8.1). This number is actually significantly higher, as the data for five West African countries is not represented here. This data also did not take into account secondary plastic forms, i.e., plastic components of non-plastic products such as electronics and vehicles, which also contribute significantly to plastic imports in West Africa. For example, in Nigeria, these secondary plastic forms made up 24% of total plastic imports in the country.

West Africa is neither among the top importers nor top producers of plastic globally (Barrowclough et al., 2020; UNEP, 2021), so these figures are paltry compared to the rest of the world. However, they are still a cause of concern as they

TABLE 8.1 Plastic imports to West African countries between 1994 and 2017 (adapted from Babayemi et al., 2019)

Country	Import Period	Plastic Imports (Tonnes)
Benin	1998–2016	386,293
Burkina Faso	1995–2016	476,548
Cape Verde	1997–2017	118,651
Gambia	1995–2016	120,433
Ghana	1996–2017	3,209,048
Guinea	1995–2015	328,744
Mali	1996–2017	496,922
Mauritania	2000–2017	477,348
Niger	1995–2016	182,811
Nigeria	1996–2017	19,865,593
Senegal	1996–2017	1,599,882
Togo	1994–2017	705,755

have only been on a steady increase in the past years. In Nigeria, West Africa's largest country, plastic waste makes up 13% of all municipal solid waste in the country (Babayemi, 2019), but though the per capita plastic consumption remains relatively low, it is predicted that the imports of plastic will double by 2030 (Aligbe, 2021).

Several factors may be contributing to this, including lack of potable water in many West African homes, leading people to heavily depend on drinking water from sachets and bottles. In Senegal, an international campaigning network reported that plastic cups and water sachets constituted almost 90% of plastic waste captured during their cleanups (Traoré, 2020). Also, as many global brands expand to Africa, plastic packaging and products have soared, with no corresponding action or infrastructure to adequately manage them. It is interesting to note that the largest plastic polluter in Africa, as identified in a global audit by an organisation fighting against plastic pollution, is not an African brand (Greenpeace, 2019). More recently, the Covid-19 pandemic meant that there was an increase in the use of personal protective equipment (PPE), most of which were largely plastic and designed to be single use. Also, lockdowns meant that e-commerce was on a rise, which led to a corresponding increase in plastic packaging, much higher than what would ordinarily be used in physical stores. The menace of plastic waste to the region has been a subject of discourse in several fora, and several campaigns against the use of plastic have been witnessed.

This "plastic pandemic" had its beginnings less than a century ago. Plastics were only introduced to Africa in the late 1950s (Jambeck et al., 2018), and despite the great advancements that have been made in science and technology, the management of plastic waste is still a current issue. In a bid to address this problem, Africa has become the "world champion" in plastic bans, leading any other continent in the number of plastic bans (Parker, 2019). In West Africa specifically, over two thirds of the countries have issued bans on plastics in some form. Though much remains to be seen on the enforcement and effectiveness of these laws, the fact remains that the governments are well aware of the plastic crisis (Greenpeace Africa, 2020).

Several factors make the management of plastic waste challenging, such as the fact that the pricing, accessibility and quality of virgin plastic are better than recycled plastic, and unlike items like glass, the functional quality of plastic greatly diminishes after recycling. There is a general lack of urgency when it comes to plastic waste, likely because people do not see how the mismanagement of plastic waste is linked to certain day-to-day problems they face such as flooding and the increased spread of diseases. Also, in a region where extreme poverty seems to be on the rise for most inhabitants (UNECA, 2021), managing plastic waste just isn't the top priority for many.

Properly managing plastic waste is an integral part of the transition towards a CPE, and technology, particularly digital innovations, may be able to help not only with the challenges faced in the general rethinking towards plastic circularity but also with the factors that elicit this great dependence on plastic. As seen in several other sectors in West Africa such as financial services and e-commerce, digital technologies have massive potential towards causing real change and innovation in the way people do things. We will proceed to view some of these cases and identify lessons and key insights that can be applied towards making plastic circularity a reality in West Africa.

2.2 Plastic Waste Management in West Africa

Plastic waste management has attracted several attentions in West Africa, but the impact seen is not tangible enough as compared to the extent of the plastic menace in the region. A number of organisations in West Africa such as Mindful Intelligent Recycling Assistant (MIRA), Dispose Green, rePATRN, Kaltani, Soso, Wecyclers, GreenHill Recycling, Trashmonger, ComeRecycle, Techbionics Ventures, Scrapays, RecycleGarb, Recycle points, Recyclan, OkwuEco, eTrash2Cash, Ecofuture, Chanja Datti, WasteBazaar, GIVO, Récuplast, Veolia, Coliba, Reaval Uno and EazyWaste have demonstrated their capacity in managing plastic waste in one way or the other. Majority of these organisations focus on collection and recycling of plastic wastes. Despite the efforts being made, these organisations are limited in their operations and impact due to lack of access to technology that can simplify their processes and ensure adequate automation of their activities. In fact, the majority of the plastic waste management companies are still manual in their operations.

Manual collection of plastic waste through human collectors is the most rampant form of collection in the region. Most households and organisations in the region do not separate their wastes, thus compounding the problem of collection of plastic wastes. In addition, the indiscipline in consumer behaviour noticeable in the act of throwing plastic bottles and sachets on the road or on the ground in public spaces also contribute to the problem of separation of plastic wastes as cleaners would just pack all the different kind of wastes and dirt together into a single bin when cleaning the roads and the public spaces. It is often observed that human collectors at dumpsites try to pick valuable plastic wastes and other recyclables from piles of mixed wastes at the dumpsites. Of course, it would be very difficult to fully separate plastic wastes from other wastes when they have already been mixed up. In fact, such plastic wastes would have little or no value as they would have been stained and made dirtier by other wastes and dirt thereby increasing the cost of preparing them for recycling. This is one of the reasons why the majority of plastic waste generated in the region is not being recycled.

After plastic wastes have been separated from other forms of wastes, sorting the plastic wastes into different types of plastic is also done manually in the region. This is time wasting and labour intensive. Plastic wastes need to be sorted before they can be optimised for recycling, reuse, upcycling or repurposing. These manual approaches cannot produce the speed and the coverage needed in transitioning to a CPE. Technologies like computer vision may be useful in auto-sorting of plastic wastes. In general, adopting technology, especially digital technologies, could enhance the transitioning to a CPE in West Africa (Oyinlola et al, 2022, 2022b).

2.3 *Digital Innovation for Plastic Waste Management in West Africa*

The technology space in West Africa received a boost within the past 10 years when the concept of innovation hubs was introduced to the region. Several innovative young minds began to team up to develop solutions to myriads of problems facing the continent. While some solutions were inspired by economic gain, several other solutions were born from the need to make social and environmental impact. As it has been widely reported, a lot of digital innovations came to being in the financial sector and have transformed payments and financial transactions in West Africa especially in Nigeria. One of the start-ups resulting from this wave is Flutterwave which has grown to become a unicorn. There is no doubt that the importance of digital innovation is well appreciated in West Africa as we see an increasing rise in the number of start-ups using digital innovations to create new experiences across several domains in the region.

Several innovators in West Africa have also ventured into solving the plastic waste problem. Table 8.2 presents some of the digital innovators, their operations and the digital technology used.

TABLE 8.2 Digital innovators in West Africa contributing to the CPE

Name	Country	What They Do	Digital Solution
Okwueco	Nigeria	Uses a mobile app as a platform to enable waste merchants to connect with sellers. The app has some other feature to make transactions seamless. The app uses image recognition to educate households about recycling and links them with merchants who can trade their waste for cash credits or mobile data transacted through the security of an online platform. Inbuilt global positioning system (GPS) facilitates logistics connecting merchants to households.	Web and mobile application
Wecyclers	Nigeria	They serve as a middleman to help in waste management control. They use mobile applications for creating awareness on plastic waste recycling.	Mobile application
Scrappays	Nigeria	Scrappay serves as a middleman in the waste management process as they connect riders (users who pack up waste) to users who need their waste to be disposed. Users can either sign up as an agent or a normal user. They also provide services to organisations.	Web and mobile applications; they also use SMS technology when there is no internet.

(Continued)

TABLE 8.2 (Continued)

Name	Country	What They Do	Digital Solution
Appcycler	Ghana	The main focus is to keep waste out of the environment. Users can schedule a pick on their website and also purchase recycled materials.	Web application
Dispose Green	Ghana	They make use of a mobile app to match the general public with waste collectors. They also keep track of the recycling process.	Mobile application
Precious plastic Gambia	Gambia	Their service is such that they provide necessities for anyone who wants to partake in waste management control. They are like a community management platform that teach people how to go about waste management control.	Mobile application
Detches a L'Or	Guinea	They provide waste technology to independent operators via a web and a mobile app. They give waste collectors the tools to ensure better delivery of service.	Web and mobile application
Pakam	Nigeria	Pakam just like others serve as a middleman between collectors (those who collect waste) and generators (those who want their waste disposed). They are currently functional only in Lagos, Nigeria.	Mobile application

3 Methodology

We employ a blended method of focus groups and one-on-one semi-structured interviews in order to gather qualitative data from CPE stakeholders in West Africa. First, a stakeholder mapping was carried out to identify stakeholders and categorise them in order to help guide the research strategy. This was accomplished by conducting desk research, obtaining referrals and consulting grey literature and online sources to discover key players in this space. The identified stakeholders were categorised based on their areas of focus into the following groups:

1. Digital Innovators/Start-ups
2. Waste Management Organisations (WMOs)
3. Civil Society
4. Academia
5. Community

3.1 Data Collection

Qualitative data was collected in two methods: First, through focus groups for three of the identified groups (WMOs, civil society and academia) and, second, through in-depth interviews for digital innovators. This was done to gain thorough comprehension of the digital innovators and start-ups on a personal level, so as to identify common themes and challenges. Participants were contacted by telephone, after which official emails containing relevant information were sent to all who indicated interest. Digital innovators identified to have contributed significantly to CPE in Africa were also invited via email and called to the interviews. Online questionnaires were also used to supplement our data collection efforts.

3.1.1 Focus Groups

The focus group is a qualitative research method used to explore the knowledge and opinions of a small set of individuals with regard to a particular topic. More viewpoints than could be gathered from the researcher alone are contributed by key stakeholders individually and collectively. The focus group meetings were preceded with an overview of the objectives of the discussion, after which pertinent consent was obtained. Following that, participants were given the opportunity to introduce themselves and their positions and work in the industry.

A series of preplanned questions were then posed to participants in order to get their informed opinions. This resulted in free-form, open-ended dialogues that influenced our research. These focus groups were very effective in capturing and taking into account the participants' emotions, expressions and opinions on the CPE in West Africa.

Focus group 1 brought together five identified WMOs within the region, while focus groups 2 and 3 brought three stakeholders in academia and civil society, respectively. Focus group 4 brought in delegates from each of the earlier focus groups in order to evoke a richer discussion on the CPE from people representing the different groups. The focus groups were held on Zoom, recorded and transcribed.

3.1.2 Interviews

Interviews are a useful technique for gathering data that involves two or more persons exchanging information through a sequence of questions and answers. The identified digital innovators were invited to one-on-one interviews to elicit rich qualitative in-depth information on the intersections between technology and the CPE. The same set of open-ended questions was posed to all participants. At the same time, additional questions were asked during interviews to clarify and/or further expand on responses given. Interviews were performed online via videoconferencing. All interviews were recorded and transcribed after obtaining the necessary consent from the participants.

3.1.3 Online Surveys

An online survey was conducted to identify the most relevant challenges affecting the transitioning to a CPE. The challenges listed were obtained from the output of the focus group discussions and the interviews conducted. The question "what do you think is the most significant challenge for sustainably managing plastic waste in your area, region, country or sector?" was posed to the respondents with six challenges highlighted, and respondents were to rate each challenge using a Likert scale of 1–7 where 1 signifies most significant and 7 signifies least significant. The six challenges highlighted are as follows:

1. Regulation (implementing and/or enforcing relevant policies),
2. Funding (accessibility to grants for research and development of innovations),
3. Awareness (education and sensitisation of the public on the impact of plastic pollution),
4. Collaboration (greater engagement between stakeholders),
5. Data (develop systems and/or technologies for capturing and tracking statistics on plastic waste) and
6. Alternative (develop alternative materials to plastics).

The respondents had the option of indicating other challenges they identified that are not part of the six.

3.2 Why This Approach

Limited structured information exists on the CPE in Africa, or West Africa in particular, so we had to make use of innovative techniques to gather data. By employing focus groups in conjunction with individual interviews and online questionnaires, we were able to gather substantial information on the current state of CPE in West Africa in a short time. There are fewer better ways to gather information on the current challenges and pain points unique to African players in this space, which was one primary focus of our research, than by hearing from the proverbial "horses' mouth" or key stakeholders themselves. We were able to generate insights grounded in the actual experiences of participants which would have been less accessible without the interaction found in a group.

We also got to see the lingo and concepts used by these groups, and the homogeneity of groups gave rise to a greater depth on certain angles and themes. From what we gathered, the focus groups also posed immense benefits to the participants. It was empowering and gave them a chance to feel seen and heard and to be regarded as authorities in their fields. Through these discussions, we got a good understanding of the current challenges faced in the transition to a CPE in Africa, and how digital innovations can help mitigate these challenges.

4 Results and Discussion

4.1 Results

Several findings emerged from the conversations had during this study. While some of the findings are new, the majority of them corroborate the existing identified challenges in the literature.

4.1.1 Focus Group 1 – WMOs

Discussion with the WMOs shows that the recyclers and collectors are already playing a big role in the plastic waste management effort in West Africa, but there was a call for the producers to wake up to their responsibility and do more. It was put forward that policies to address this gap need to be developed and effected. Apart from the producers, organisations that generate a lot of plastic waste need to be more involved in the drive to a CPE. The need for a multilayer collaboration among the government, businesses and societies in the CPE drive was pointed out. Another major theme emanating from the conversation was the use of incentives to encourage people to be involved in plastic waste management. Various examples of where this has been used were given. It was also noted that a lot of data on plastic waste management was not available, and this needs to be worked on. It was also identified that a major problem faced by the WMOs was that of logistics. Some solutions like alert systems and apps were suggested. Another problem was the exorbitant fines the organisations had been charged just by carrying out their activities. One suggestion is to provide tax holidays to these organizations, while another is to conduct a comprehensive review and provide clear explanations of policies for better understanding. It was also recommended that the independent scavengers and recyclers need to be well integrated into a system that works.

4.1.2 Focus Group 2 – Academia

In this discussion, participants highlighted some gaps in pursuing research in the CPE, which included a lack of funding or more appropriately lack of awareness of funding opportunities as well as the unwillingness of plastic-producing companies to embrace alternatives. It was stated that these gaps could be closed with the existence of policies to encourage plastic producers to support research, digital innovations to aid town-and-gown interactions and a need for researchers from different disciplines and institutions to work together. A glaring problem observed was that despite the large amount of plastic waste being generated in West Africa, a shortage of materials to recycle exists which implies that inability to properly collect and sort plastic wastes makes it impossible to take advantage of generated waste for recycling purposes; hence, the majority of the waste ends up in landfills and the ocean.

4.1.3 Focus Group 3 – Civil Societies

The need for more engagement and advocacy in rural areas was underscored, as well as the urgency to transition from the current model of sporadic cleanups to a more sustainable plastic waste management culture. It was also emphasised that advocacy work and consumer education needed to be targeted towards the youth, as they make up most of the population. Following from this, participants delved into explaining how changes in plastic design, to enable them to be safely reused for the same purpose (particularly food-grade products), as opposed to being downgraded in use, will be highly encouraged. Also, for other plastic products, participants said that if plastic waste could be proven to be converted to commercially viable products, it will lead to more income streams and employment opportunities for the populace. Lastly, participants highlighted the need for increased collaboration among several players in order to cover more ground in this transition.

4.1.4 Online Survey

A total of 39 stakeholders responded to our survey in ranking the major challenges identified based on the conversation they had during the focus group discussion and the interviews conducted. Figure 8.2 shows the relevance of the challenges. Regulation was considered the most significant as it was ranked as number 1 by 29 respondents. This was followed by awareness ranked as number 1 by 28 respondents. Funding, collaboration, data and alternative product were ranked by 19, 18, 16 and 13 respondents, respectively, as the most significant challenges.

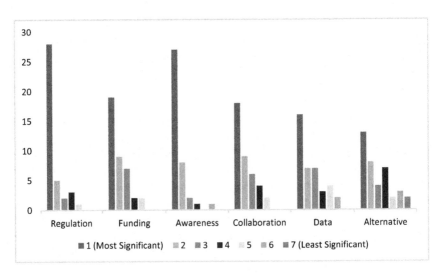

FIGURE 8.2 Ranking of challenges by stakeholders

4.2 Discussion

Our findings in this research show that several initiatives are ongoing in West Africa to transition to a CPE; however, the number and impact of the initiatives do not match the volume of usage of plastic products and the impact of plastic waste generated both locally and from foreign imports. Every participant in the research agreed on the negative impact of plastic to our environment and also on the need to move towards a CPE. The major challenges observed from our findings that could be addressed using digital innovations are discussed below.

4.2.1 Challenge I – Awareness

We observed that consumer behaviour (both at individual and organisation levels) is one of the biggest challenges facing CPE transitioning in West Africa. This is due to the fact that users are under-aware of their usage actions and also because there are no immediate consequences or penalties accruing to users. This brings awareness to the front burner as one of the major challenges to be tackled if the dream of transitioning to a CPE in the region must be achieved. Below are some quotes from the conversation that highlights this:

> We know that there is a gap in terms of the advocacy, education, and awareness. If not, people will not be dumping their packaging materials in the environment.
>
> Even if there's no recycling facility in Nigeria, it shouldn't be in the environment. It should be collected separately.
>
> I think that the most important role here is on the consumer because consumer can hold the producer to account and say 'if you don't consider the environment, I'm just not going to buy from you'.
>
> So those at the rural level, probably are thinking about how to eat three times a day and how to have those basic necessities of life. How can you communicate to them that they're endangering the environment by their consumption patterns? It just wouldn't add up or makes sense. So, it's all about innovation in terms of passing the message across. Just making them understand that they are endangering themselves indirectly by these unsustainable practices. We need to have translators and local dialects also.

There is a need to design a whole multilayer, status-targeted, culture-specific and social-oriented awareness framework that can address this cancerous challenge. Creative digital innovations could be developed to address the awareness challenge. Currently, most of the digital innovations on CPE in the region do not address the awareness problem. Until the awareness level is raised high and permeated through citizen education across the entire region and across the different cultures and social statuses, efforts in transitioning to a CPE might not yield a satisfactory result. As part of the efforts to solve the awareness challenge, the government must come up with regulation (policies) that penalises consumers for improper disposal of waste. Such regulation and the penalty for breaking such should be a component of the awareness programme design. Penalty alone will not solve the problem; the government in partnership with the private sector should also provide efficient waste collection services that the consumers can subscribe to for proper disposal of their waste. In addition, public spaces should have dedicated waste bins for plastic wastes in order to facilitate separation of waste at source.

4.2.2   Challenge II – Data

Another challenge plaguing the transitioning to a CPE in West Africa is the problem related to generating accurate data on plastic life cycle in the region. Inability to monitor plastic from production to disposal is a problem that affects decision-making around plastic waste and effective optimisation of the recycling, upcycling and repurposing processes. We captured this challenge using the words of the participants in the following quotes:

> You would find a study that says a researcher says 13,000 metric tonnes are being generated daily or 10,000 tonnes are being generated daily. So different researches have different numbers and it's a bit difficult for us to categorically say, oh, this is the percentage that we've been able to manage.
>
> I think it's quite important to have these numbers because it dictates the right level of digital innovation needed to help. If you want to design a tool to facilitate operations, you need to know about the extent and the magnitude of the waste generated.
>
> An app that can help aggregate all the information in one place will basically just help the easy spread of information.

A data-driven decision-making process on plastic waste management and plastic consumption regulation can fast-track the transitioning to a CPE. Creating digital platforms for data logging, data aggregation and data visualisation on plastic production, distribution and waste, and monitoring the life cycle of every plastic product from production to disposal could boost CPE transitioning efforts and provide insight that innovators or start-ups need for new ideation output on technologies to support the CPE transitioning.

4.2.3   Challenge III – Logistics

From our discussions with the different stakeholders, we discovered that logistics around plastic waste collection and mobility is another big challenge as it was observed that the cost of logistics around mobility of plastic waste is high as compared to the economic value of the waste. This is also related to the problem of awareness because separation of plastic waste is not done at source. If wastes in general and specifically plastic waste are separated from source, it will be difficult to see them littering the community, and it would make collection easier for collectors and facilitate proper planning and optimisation of the collection process.

> The logistics is a nightmare. So, it would be nice if someone just told me that we could actually outsource our logistics and assign a percentage of our monies to it and let the people pick up and drop off at the location.
>
> So, there are a whole lot of challenges around the pickup and logistics like vehicles breaking down … I think most of the participants here would relate to that with regards to their vehicles, with regards to paying different kind of bills and you don't even know where those bills are being generated from.
>
> For normal municipal waste collections, you'll see that having those sorts of automated optical sensors that can tell you when the bin is full or it is half full will enable most of the collection to be done seamlessly. Waste will not be allowed to stay in one place over a long period of time.
>
> Even the logistics of picking it up was even more expensive than what I was collecting. So, it's quite frustrating doing this business in Nigeria.

Well thought-off and innovative digital solutions are needed that can help in optimising logistics around mobility of plastic waste.

4.2.4 Challenge IV – Lack of Competence in New Digital Technologies

In terms of using digital innovation in solving the plastic problem, our findings show that most of the digital innovators in West Africa use the same technology, majorly mobile apps and web apps, for communication and workflow among the waste collectors, aggregators and recyclers. It was also obvious that the digital innovators in the region lack competence in integrating new technologies like artificial intelligence (AI), internet of things (IoT), blockchain, Extended Reality (XR), etc. in their innovations. This resonates well with our findings in another work detailing the tech readiness of digital innovators in Africa (Kolade et al., 2022) (Figure 8.3).

4.3 Case Studies

We present two case studies of digital innovators in West Africa that are key players in enhancing circularity in the region.

4.3.1 GIVO

4.3.1.1 Background and History

Founded in 2019, GIVO (Garbage In, Value Out) is a Nigerian start-up that leverages technology to collect recyclable plastic directly from individuals,

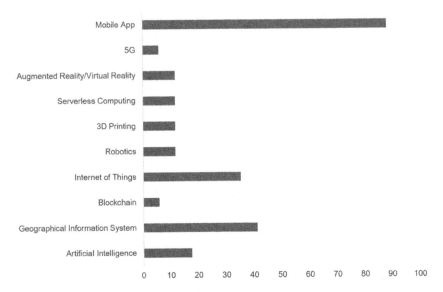

FIGURE 8.3 Tech readiness index of the DIs in each of the ten frontier digital technologies (Kolade et al., 2022)

households and businesses and then process the collected plastics into finished products. GIVO started in response to the Coca-Cola Entrepreneurs Plastics Innovation Challenge (EPIC) and has since grown to collect plastic waste which is then processed into consumer and industrial goods.

4.3.1.2 Innovation and Business Model

GIVO automates and digitises the polyethylene terephthalate (PET) recovery process using an IoT, app and web-based solution. Plastic waste is collected and processed through modular recycling centres established all around the country. Once collected, the plastic materials are measured, weighed, washed, resized and catalogued using IoT technology before being shredded. The GIVO set, which does this, contains software and hardware (a scale, an android device, a printer, a shredder and a solar power solution all enabled by IoT).

Each GIVO centre, housed in a portable 40-ft container, is managed by a franchisee. Customers or scavengers in the neighbourhood either bring their raw PET to the franchisee or ask for the nearest franchisee to pick it up. The franchisee identifies the buyer (using the android) and weighs the products (using the IoT scale). A receipt for the transaction is generated for the customer (using the IoT printer) once the product's worth has been verified. The franchisee then gathers all the PET that was collected per day, sorts it by type and then feeds it into the (IoT) shredder to make a new product (shredded PET).

When the amount of collected shredded PET reaches a certain amount (say 100 kg), an automatic request for pickup is made to the aggregator with information on the amount, price and location of the shredded PET. Using an IoT scale and an android device, the aggregator selects the goods, confirms its value and pays GIVO. (This is shared 25:75 between the GIVO platform and the GIVO franchisee.)

Every step of this process, including identifying stakeholders, accepting payments, organising logistics and weighing products, is entirely digitalised and accessible through a dedicated back end. The recyclable plastics are then used to make high-quality products for their customers. In the heat of the Covid-19 pandemic, GIVO was producing international standard PPE, including face shields and masks, as well as flowerpots, toys and Christmas ornaments. They have made over 15,000 units of the products. They also produce "Eco Panels", which are made from one plastic type (100% recycled) and can be used for construction purposes. They can also be employed to manufacture furniture including tables, kitchen cabinets, countertops, etc.

In the Nigerian CPE, record keeping, data gathering and analysis are manual and fairly inconsistent. Also, the available data usually rests in the hands of a few and is not publicly available. With their solution, GIVO is gathering data accurately, getting insights from this to optimise processes and democratising access to this data. The data collected is able to help with carbon credits and provides information on the brands commonly used by customers in the area. GIVO is also able to give real-time info on brand positioning and brand behaviour in certain areas with this data. IoT and AI are used to digitise their collection process, track recyclables collected and processed, optimise operations and generate data on waste consumption patterns within communities.

4.3.1.3 Impact

Due to heavy reliance on imported goods, the manufacturing sector in Nigeria currently contributes about 9% to total GDP (National Bureau of Statistics, 2021). GIVO attempts to solve this problem by exploiting the raw plastic waste materials generated in the country to locally manufacture and meet the needs of the population.

Each GIVO centre is community based and led by women and children. Team members from these community centres regularly go out to educate inhabitants on the benefits of proper plastic waste management. In each community, each centre employs between 10 and 15 staff. Through the centres, GIVO is able to provide formalised jobs and promote financial inclusion through incentives for women and youth. They also encourage entrepreneurship in these communities by operating a franchise model for the centres.

These centres rely on solar power, reducing the carbon footprint of their solution. In just 18 months since they launched, they had been able to collect plastic waste from about 200 households in over 700 unique transactions.

In partnership with the Lagos State Waste Management Agency (LAWMA), the company conducted a pilot with a recycling supplier to digitise their collection processes and provide a deeper insight of recycler activities in Lagos State. One hundred and fifty-seven households and companies signed up for the 6-week experiment, which resulted in 674 collections and the collection of over 1452 kg of recyclable plastic.

Now, according to them, they collect about 300 kg of plastic waste per day, which translates to 7 metric tonnes a month and 90 metric tonnes per year. This will reduce carbon emission by 540 Mt annually. The business has forged partnerships with organisations including WestAfricaENRG, Orange Corners, AFD, FATE Foundation and others.

4.3.1.4 Challenges

A major challenge faced by the company is proper education of the entire populace in order to drive the recycling culture. Also, GIVO hopes to translate their model to other recyclers in the region. However, the majority of recycling businesses use manual procedures and operate on extremely slim margins. As a solutions provider, the majority of the recyclers cannot afford to adopt their solutions.

4.3.2 Coliba

4.3.2.1 Background and History

Coliba Ghana puts technology at the centre of their plastic recycling operations by utilising web, mobile application and SMS to facilitate the collection of recyclables. They connect households and businesses with waste pickers, give people the opportunity to monetise their plastic waste products and provide a steady supply for recycling companies. Coliba started in 2016 after the founder lost a friend in the June 2015 Accra floods which are believed to have been caused or exacerbated by plastic pollution (Raymond, 2015). A pilot was carried out in five schools in Ghana after which operations began fully.

4.3.2.2 Innovation and Business Model

Coliba operates a franchise model through its multi-sided digital platform where people who handle plastic waste pickups, i.e., "Coliba Rangers", can register to discover locations for pickups, and individuals and institutions can register to find the nearest "Ranger" to pick up their recyclables.

With geolocation features, users can find and request a ranger and schedule a pickup. The ranger will then pick up the waste and deliver them to the recycling parks for further processing. In addition to the application, rangers can be reached through instant messaging platforms like WhatsApp. The app also serves as a tool to educate users on effective plastic waste processing practices, such as sorting and separation.

Coliba is also plastic credit certified, which means companies can purchase plastic credits from them to meet their plastic waste reduction or plastic–neutrality goals. A plastic credit is a transferable certificate that is issued when a specified quantity of plastic waste (kilograms, metric tonnes, etc.) that would otherwise have polluted the environment is recovered and properly managed. Monthly audits are conducted for them to accurately trace the sources of plastic waste and gain better understanding of the entire plastic value chain.

People can also deposit plastic waste products to Coliba's community buy-back centres, mostly run by women, in exchange for incentives in cash and kind. The recovered plastic is then taken to the Coliba recycling park, where it is separated and recycled into pellets and flakes (including high-density polyethylene (HDPE) and low-density polyethylene (LDPE), and later sold or exported to customers in order to create new products for the local and international markets.

Through the hubs and centres, they are able to create jobs and help the planet, thereby championing the circular economy model. Coliba also has solar-powered office spaces created entirely from reused shipping containers.

To achieve their mandate, Coliba leverages partnerships. For example, Coliba recycling bins are placed at select filling stations around the country and serve as a source for recyclable plastic. They also partner with food brands, insurance companies and schools, and people can exchange their plastic products for meal tickets, insurance, school fees, phone credit and vouchers. They have also received funding from the Dutch Embassy in Ghana and United Nations Development Programme (UNDP).

4.3.2.3   Impact

Since operations began, the company has been able to set up about 40 recycling centres across Ghana, thereby retrieving over 700 metric tonnes of plastic waste which corresponds to a carbon offset of 4200 tonnes. They also organise ocean cleanups of water bodies and leverage social media to gather volunteers. Over 13 metric tonnes of plastic can be gathered in a single cleanup, and this is done several times per year.

The community buy-back centres and recycling parks have employed at least 310 workers, 80% of whom are women. They have been able to better the livelihoods of their waste pickers, who are mostly widows and single mothers. The waste pickers are trained in environmental sustainability and waste processing, giving them skills that better position them economically. Rural communities also receive support in terms of education and insurance as incentives to drop off their plastic waste.

4.3.2.4   Challenges

Currently, the majority of the recycled plastic used for manufacturing are of low quality, which means they may not be able to be recycled more than once. Funding is a major challenge to solving this problem as recycling is very capital intensive, more so when the products are to be made at very high quality. There is a need to establish more collection centres and purchase better quality machinery, in order to keep plastics in the processing cycle for longer and achieve a truly circular plastic value chain.

There is also a lack of regulations and enforcement on plastic waste management, and this can complicate work for waste pickers. Lack of awareness on sustainable production and consumption is also a challenge the company faces when dealing with potential customers and partners, and more targeted education for the populace and extended producer responsibility policies can help mitigate this.

5   Conclusion

Despite the environmental and economic gains that can be achieved through a CPE, the transition to a CPE in West Africa is plagued with several challenges. Paramount among the challenges as identified above are issues relating to awareness, policy, data, logistics and lack of competence in frontier digital technologies. Even though policy was identified as a challenge, the problem isn't the lack of policy per se but rather the poor or zero implementation and enforcement of established policies (Schroeder et al., 2023). This nonetheless does not negate the need for more and better policy formulation by the government. Overall, awareness seems to be the biggest challenge in the sense that it is the foundation that is needed in solving all other challenges. Awareness will greatly influence consumer behaviours and also change the disposition of plastic producers. Awareness of the health and environmental impact of plastic wastes through campaigns from civil societies can also push the government to formulate better policies, develop more regulations to address the plastic issue and put stricter measures in place to enforce compliance.

In addressing the awareness problem, we need to expand the concept of plastic pollution well beyond ceremonial cleanups in an effort to influence the attitudes and practices of major consumers of plastic, such as businesses, universities and hospitals.

Adopting existing digital innovations and creating new ones can help in the awareness drive. Internet and social media can greatly facilitate this. This will be particularly useful among the youth, who make up most of the population and can also be targeted by leveraging these platforms. Many youths increasingly say social media platforms serve as their sources for information on issues ranging from social justice to news (Suciu, 2022). Youth also leverage these media as tools for expression, citing their views on issues they consider important (Booth, 2021). An example of when social media was used as a tool for action was with the ENDSARS protests in Nigeria in 2020, where the vast majority of Nigerian youth took to the streets (online and offline) to protest against police brutality.

In a survey carried out by Reach3 insights exploring the role of "Tiktok" – a video-sharing app that allows users to create and share short-form videos – on social activism, climate change emerged as one of the top ten things the younger generation (age 13–24) cared about. Fifty-four percentage of respondents said that they have engaged in discussions with friends and family because of content they saw on the platform, 44% have signed petitions and 32% educated themselves further due to what they saw on the platform (Hosie, 2020). This indicates that leveraging social media could be the direction to go in increasing awareness of the CPE in West Africa. An example of this is the UNEP campaign, *#BeatPlasticPollution*, launched in 2017 to encourage people to adopt one single behaviour for 100 days, in order to reduce pollution caused by plastics. Among other things, participants in the campaign were encouraged to give up/reduce the use of single-use plastics, raise funds and organise events for more sensitisation worldwide.

Furthermore, the use of incentives, though unsustainable, can be a good way to get more people to be aware of the impacts of their actions in the transition to a CPE and can ultimately drive behavioural change. For example, if customers pay a deposit, which is then repaid when plastic bottles are returned, the low rate of plastic collection in Africa might be drastically improved. Deposits are already required for glass bottles in some parts of the continent. One of the participants in the focus group said

> We started the plastic-waste-for-income-model after the COVID19 lockdown and so far, we have over 150 subscribers within the community. Many use it as an alternative source of income. They use it to pay for things like data subscription and TV streaming services. Some even use it to get books.

At the DITChPlastic Hackathon held in Lagos, one of the winning solutions was a website where users can register and then submit recovered plastics to established collection hubs. In return, they get coins added to their profile on the site. Coins accumulated to a certain amount can then be used to redeem gifts like foodstuffs. Blockchain can also be used to accomplish this. For example, The Plastic Bank Recycling Corporation, offers blockchain-secured digital tokens in exchange for recycled plastic (Katz, 2019). The United Nations Framework Convention on Climate Change also launched a Climate Chain Coalition in 2018 where over 80 groups have pledged to use blockchain technologies to combat climate change (United Nations Climate Change, 2018). This shows that there are opportunities for African start-ups in the Web3 and blockchain sectors to build scalable business models around the CPE.

In conclusion, to sustain the gains already achieved in plastic waste management and to fast-track our transitioning to a CPE, digital innovators in West Africa need upskilling in frontier digital technologies as they continue to develop new solutions to address the challenges affecting the circularity of plastic in the region.

Acknowledgement

This work was partly supported by the United Kingdom Research and Innovation (UKRI) Global Challenges Research Fund (GCRF) under Grant EP/T029846/1.

References

Acquah, R. (2015). The 2015 June 3rd Twin-Disaster in Accra: A Situational Analysis of Ghana's Disaster Preparedness, University of Agder Master's theses in Global Development and Planning.

Aligbe, M. O. (2021). Investigating the Use of Plastic Bags in Lagos, Nigeria (Dissertation). https://doi.org/2021/8

Babayemi, J. O., Nnorom, I. C., Osibanjo, O., Weber, R. (2019). Ensuring sustainability in plastics use in Africa: Consumption, waste generation, and projections. *Environmental Sciences Europe*, 31(1). doi:10.1186/s12302-019-0254-5

Barrowclough, D., Birkbeck, C. D., Christen, J. (2020). Global Trade in Plastics: Insights from the First Life-Cycle Trade Database, UNCTAD Research Paper No. 53 UNCTAD/SER.RP/20200/12. Available at: https://unctad.org/system/files/official-document/ser-rp-2020d12_en.pdf

Berg, N. (2012). With Urbanization Comes Mountains of Trash. Bloomberg Africa Edition. Available at: www.bloomberg.com/news/articles/2012-06-13/with-urban ization-comes-mountains-of-trash

Booth, Ruby Belle (2021). Young People Created Media to Uplift Their Voices in 2020. Tufts University Center for Information & Research on Civic Learning and Engagement (CIRCLE). Available from: https://circle.tufts.edu/latest-research/young-people-crea ted-media-uplift-their-voices-2020

Greenpeace (2019). Branded – Identifying the World's Top Corporate Plastic Polluters. Vol II. [Online] Available from: www.breakfreefromplastic.org/wp-content/uploads/2020/07/branded-2019.pdf

Greenpeace Africa (2020). 34 Plastic Bans in Africa | A Reality Check [Online] Available from: www.greenpeace.org/africa/en/blogs/11156/34-plastic-bans-in-africa/

Hosie, Katherine (2020). More than just Tok: Gen Z's Activism on TikTok is Outperforming the Performative. *Reach3 Insights*. Available from: www.reach3insights.com/blog/tik tok-social-activism

Jambeck, J., Hardesty, B.D., Brooks, A.L., Friend, T., Teleki, K., Fabres, J., Beaudoin, Y., Bamba, A., Francis, J., Ribbink, A.J., Baleta, T., Bouwman, H., Knox, J., Wilcox, C. (2018). Challenges and emerging solutions to the land-based plastic waste issue in Africa. *Marine Policy*, 96, pp. 256–263.

Kariba, F. (2020). The Burgeoning Africa Youth Population: Potential or Challenge? Cities Alliance [Online]. Available at: www.citiesalliance.org/newsroom/news/cit ies-alliance-news/%C2%A0burgeoning-africa-youth-population-potential-or-challe nge%C2%A0 (Accessed: 2 Dec 2022).

Katz, David. (2019). Plastic Bank: Launching Social Plastic® Revolution. Field Actions Science Reports, Special Issue 19 | 2019, 96–99.

Kaza, Silpa, Yao, Lisa C., Bhada-Tata, Perinaz, Van Woerden, Frank. (2018). What a Waste 2.0: A Global Snapshot of Solid Waste Management to 2050. Urban Development. Washington, DC: World Bank. © World Bank. https://openknowledge.worldbank. org/handle/10986/30317. License: CC BY 3.0 IGO.

Kolade, O., Odumuyiwa, V., Abolfathi, S., Schröder, P., Wakunuma, K., Akanmu, I., Whitehead, T., Tijani, B., Oyinlola, M. (2022). Technology acceptance and readiness of stakeholders for transitioning to a circular plastic economy in Africa. *Technological Forecasting and Social Change*, 183, 121954.

Mavropoulos, A., Mavropoulos, A., Koukosia, I., Tsakona. M., Mavropoulou, N., Rigas, N., Andreadakis, T. (2014). The World's 50 Biggest Dumpsites, Waste Atlas 2014 Report.

Merem, E.C., Twumasi, Y.A., Wesley, J., Olagbegi, D., Crisler, M., Romorno, C., Alsarari, M., Isokpehi, P., Afarei, M., Ochai, G.S., Nwagboso, E., Fageir, S., Leggett, S. (2021). Analyzing the environmental risks from electronic waste dumping in the West African region. *Journal of Health Science*, 11(1), pp. 1–16.

National Bureau Of Statistics (2021) Nigerian Gross Domestic Product Report (Q1 2021)

Okafor-Yarwood, I., Adewumi, I. J. (2020). Toxic waste dumping in the Global South as a form of environmental racism: Evidence from the Gulf of Guinea. *African Studies*, 79(3), 285–304.

Oyinlola, M., Kolade, O., Schroder, P., Odumuyiwa, V., Rawn, B., Wakunuma, K., Sharifi, S., Lendelvo, S., Akanmu, I., Mtonga, R., Tijani, B., Whitehead, T., Brighty, G., Abolfathi, S. (2022b). A socio-technical perspective on transitioning to a circular plastic economy in Africa. *SSRN Electron. J.* https://doi.org/10.2139/ssrn.4332904

Oyinlola, M., Schröder, P., Whitehead, T., Kolade, S., Wakunuma, K., Sharifi, S., Rawn, B., Odumuyiwa, V., Lendelvo, S., Brighty, G., Tijani, B., Jaiyeola, T., Lindunda, L., Mtonga, R., Abolfathi, S. (2022b). Digital innovations for transitioning to circular plastic value chains in Africa. *Africa Journal of Management*, 8(1), pp. 83–108. https://doi.org/10.1080/23322373.2021.1999750

Parker, L. (2019). Plastic Bag Bans Are Spreading. But Are They Truly Effective? National Geographic [Online]. Available from: www.nationalgeographic.com/environment/article/plastic-bag-bans-kenya-to-us-reduce-pollution

Schroeder, P., Oyinlola, M., Barrie, J., Bonmwa, F., Abolfathi, S., 2023. Making policy work for Africa's circular plastics economy. *Resources, Conservation and Recycling*, 190, 106868. https://doi.org/10.1016/j.resconrec.2023.106868

Signé, L. (2019). Africa's Emerging Economies to Take the Lead In Consumer Market Growth. Brookings, Africa In Focus. Available at: www.brookings.edu/blog/africa-in-focus/2019/04/03/africas-emerging-economies-to-take-the-lead-in-consumer-market-growth/

Staatz, J., Hollinger, F. (2016). West African Food Systems and Changing Consumer Demands. West African Papers, No. 04, OECD Publishing, Paris. http://dx.doi.org/10.1787/b165522b-en

Suciu, Peter. (2022). Teens Increasingly Rely on Social Media for News – But They Don't Trust It. *Forbes*. Available from: www.forbes.com/sites/petersuciu/2022/07/22/teens-increasing-rely-on-social-media-for-newsbut-they-dont-trust-it/?sh=19b00bf419ff

Torres, C., Seters, J. (2016) Overview of Trade and Barriers to Trade in West Africa: Insights in Political Economy Dynamics, with Particular Focus on Agricultural and Food Trade. European Centre for Development Policy Management. Available at: https://www.tralac.org/images/docs/10274/overview-of-trade-and-barriers-to-trade-in-west-africa-insights-in-political-economy-dynamics-agricultural-trade-ecdpm-july-2016.pdf

Traoré, A. (2020). West Africa Breaks Free from Plastic, Greenpeace [Online]. Available from: www.greenpeace.org/africa/en/blogs/9008/west-africa-breaks-free-from-plastic-time-to-celebrate-victories/

United Nations Climate Change (2018). UN Supports Blockchain Technology for Climate Action. Available from: https://unfccc.int/news/un-supports-blockchain-technology-for-climate-action

United Nations Department of Economic and Social Affairs Population Division (2015). Youth Population Trends and Sustainable Development. Population Facts [Online]. Available at: www.un.org/esa/socdev/documents/youth/fact-sheets/YouthPOP.pdf (Accessed: 2 Dec 2022).

United Nations. Economic Commission for Africa; United Nations. World Food Programme; United Nations. Economic Commission for Africa; United Nations. World Food Programme). (2021-08). Monitoring report on the impacts of COVID-19 in West Africa. Addis Ababa: © UN. ECA, https://hdl.handle.net/10855/47581"

United Nations Environment Programme (2021). Drowning in Plastics – Marine Litter and Plastic Waste Vital Graphics. Available at: https://www.unep.org/resources/report/drowning-plastics-marine-litter-and-plastic-waste-vital-graphics

Wilson, D.C., Rodic, L., Modak, P., Soos, R., Carpintero, A., Velis, K., Iyer, M., Simonett, O. (2015). *Global Waste Management Outlook*. UNEP.

9

A MULTI-STAKEHOLDER, MULTI-SECTORAL APPROACH TO A CIRCULAR PLASTIC ECONOMY IN EASTERN AFRICA

Oluwaseun Kolade, Muyiwa Oyinlola and Barry Rawn

1 Introduction

East African countries have experienced significant economic growth in the past decades. This has, among others, precipitated a significant increase in the quantity of plastic products imported into the region (Oyake-Ombis et al., 2015). Between 2016 and 2019, the volume of plastic wastes increased by 28% in Kenya, 48% in Tanzania, 94% in Ethiopia, and 45% in Uganda (Regional Economic Department Kenya, 2022). The challenge of plastic pollution is exacerbated by societal lock-in into the linear economic habits of consumption and the inadequacy of infrastructures for management of plastic wastes. Added to this are the challenge of the institutional environment and the inadequacy of policies and regulations to effectively grapple with the growing menace of plastic wastes in the region.

The challenge of plastic waste in East Africa reflects a wider trend across the continent, where economic growth has been observed to be directly proportional to the volume of plastic wastes (Babayemi et al., 2019). Therefore, as growth continues to gather pace on the continent, the imperative of conversations about sustainability and circular economy becomes more urgent. While there are inspiring examples of innovations for the circular plastic economy on the continent, the overall picture is mixed, mainly because stakeholders continue to work in silos and therefore unable to harness the collective synergy for maximum impact (Kolade et al., 2022). In order to break the lock-in to the linear economy and accelerate the transition to the circular economy, stakeholders across public, private and the third sector must pool resources and knowledge together to develop and promote new innovations.

The East African region is undergoing structural economic transformation and growth. Following the slowdown of the economy precipitated by Covid-19

DOI: 10.4324/9781003278443-11

pandemic, East African countries are currently rebounding. This is driven by increasing movement of labour and productivity from agriculture to higher value sectors of manufacturing and services. In Tanzania, the industrial sector accounted for 0.6 of the 2.1% gross domestic product (GDP) growth in 2020, and 2.6 of the 6.1% growth for Ethiopia (African Development Bank, 2021). While national governments across East African countries are enacting policies and regulations to stem the problem of plastic waste, the results have been generally modest and mixed. Rwanda, for example, has had considerably bigger success in implementing plastic bans, compared with countries like Kenya and Uganda. Some stakeholders have argued that variations in successful implementation of policies can be explained by differences in levels of business power, given that plastic manufacturers are fewer and smaller and therefore limited in economic and political leverage in Rwanda (Behuria, 2019). Others have noted that business power is not a sufficient explanation of the variations because the local and external environments also have significant impacts on successful innovations of environmental policies.

The rest of this chapter is organised as follows: First, we provide an overview of three country contexts of Kenya, Rwanda and Uganda, to highlight key issues and peculiarities in the policy and regulatory landscape. We then describe the methodological approach, before presenting and discussing the findings from focus group discussions and in-depth interviews held with selected participants across the East African region. This chapter concludes with a summary of key insights from East Africa that can help drive the transition to a circular plastic economy in Eastern Africa.

2 Focal country contexts

The following sections provide an overview of policy and political contexts of circular plastic activities and outcomes in three focal countries in the East African region. A summary of these is presented in Figure 9.1. It is important in global conversations about the circular economy to understand the differences as well as similarities across countries. This is necessary for better policy outcomes achieved through exchange of best practices across countries and design of bespoke policies that address specific challenges and needs.

2.1 Kenya

Waste generation is generally low in Kenya, at an average of 11 kg per capita annually, compared with the global annual average of 29 kg per capita (Griffin and Karasik, 2022). However, about 92% of solid waste is mismanaged, partly due to the absence of collection facilities in rural areas and increasing leakages from urban centres. The key sectoral contributors to plastic waste are packaging, textiles and automotive tyres. In the last 10 years, waste generation (4 Mt/year)

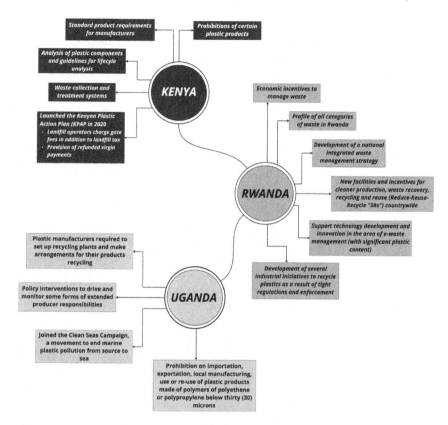

FIGURE 9.1   Overview of circular plastic policies in Kenya, Rwanda and Uganda

has increased greatly with Kenya's rapid urbanisation and is expected to double by 2030. Currently, waste management structures fail to address the magnitude of the problem. In the capital region of Nairobi, only about 20% of the solid waste (1 Mt/year) is recovered for recycling. The remaining 80% is left on the streets. Existing landfills have by far exceeded their capacities to safely dispose of the waste volumes, thereby degrading the environment and adversely affecting human health.

Kenya's policy response to the problem of plastic waste has been organised around three key areas: prohibitions of certain plastic products; standard product requirements for manufacturers; and waste collection and treatment systems (UNIDO, 2022), including recycling. Kenya is one of the few countries in Africa which has banned the use of single-use plastic bags in 2017. However, the issue of plastic waste management is associated with the general poor state of solid waste management (SWM) and the poor infrastructure. Kenya national SWM policies, environmental policy and SWM strategy are aligned to regional and global targets but currently fail to achieve them. Findings from a recent review work highlighted the need for a clear (1) coordination mechanism for policy implementation and evaluation; (2) alignment among the different provisions and synergy in their implementation; (3) enhancing institutional capacity (infrastructural, financial and human resources) of key actors in the government sector for successful implementation of the policies.

The Kenyan Bureau of Standards (KEBS) publishes and oversees the enforcement of standards requirements for the manufacture of plastic products. The standards framework includes provisions for analysis of plastic components and guidelines for life cycle analysis. Recently, regulators have given increasing attention to requirements for biodegradability and compostability of plastic products (UNIDO, 2022). In response to conversations initiated by the Kenyan government about extended producer responsibility (EPR), the Kenyan Association of Manufacturers (KAM) launched the Kenyan Plastic Action Plan (KPAP) in 2020 (UNIDO, 2022). Under KPAP, landfill operators charge gate fees in addition to landfill tax imposed by public authorities. In addition, KPAP also provides for "refunded virgin payments". Under this, producers whose products consist of mainly virgin materials pay a fee that is used to refund producers who use mainly recycled materials (KAM, 2019).

In addition to the financial elements, the KPAP also comprises six other key elements: recycling options, segregation at source and waste collection, product design for enhanced recycling, consumer awareness campaigns, biodegradable plastics and integration of the informal sector (KAM, 2019). KPAP effectively recognises the importance of a whole-value chain approach to a circular plastic economy in Kenya and East Africa. There are specific measures aimed at the design and production stage, including the financial instruments such as the refunded virgin payments. The inclusion of design for enhanced manufacturing also underlines this increasing focus on the earlier stages of the plastic value chain. The plan also highlights the importance of two categories of stakeholders that are not typically given prominence in discussions about the circular plastic economy: consumers and informal waste collectors. These groups of stakeholders are critical for successful transition to a circular plastic economy. Like producers, consumers also need incentives to embrace new, circular habits of consumption and therefore contribute to breaking the lock-in to the linear economy. Similarly, infrastructures on their own are inadequate for effective management of plastic wastes without the critical contributions of human actors, such as informal waste reclaimers, who make the infrastructures work.

2.2 Rwanda

The rapid increase of Rwanda's population has stretched the current infrastructure resulting in many complex problems regarding municipal solid waste (MSW) management. These are shared problems with other low income (LI) and low and middle income (LMI) countries such as inadequate service provision, limited recycling activities and insufficient/ineffective landfill management. In Kigali, recent estimates suggest a production of around 2 kg of waste per person, per day, with an average content of 1.5% of plastic. Although integrated waste management strategies at national and local levels are still missing, the government is taking actions against plastic pollution to promote environmental awareness and find credible solutions to eradicate plastic waste. In 2008, the country banned importation and use of polythene bags and started environmental campaigns to monitor the ban. Rwanda's example shows how decisions taken at a national level and enforced proactively can cut down on the use of plastics. The experience in Rwanda contrasts with the otherwise increasing plastic consumption in other African countries. The country therefore bucks the trend linking economic growth with increasing amounts of plastic wastes. Between 2008 and 2017, Rwanda experienced an increase in GDP per capita from $1229 per year to 2080. Roughly within the same period, between 2007 and 2016, the importation of finished plastics declined from about 2700 tonnes in 2008 to 175 tonnes in 2016. This provides a good example of sustainable, green growth.

The Rwanda's National Environment and Climate Change Policy, revised in 2019, identified seven essential objectives for achieving a sustainable and green nation including (1) the development of a national integrated waste management strategy; (2) economic incentives to manage waste; (3) new facilities and incentives for cleaner production, waste recovery, recycling and reuse (Reduce-Reuse-Recycle "3Rs") countrywide; (4) a profile of all categories of waste in Rwanda; and (5) supporting technology development and innovation in the area of e-waste management (with significant plastic content). The implementation plan aims at setting up a "profile of all categories of waste used in Rwanda" and develop an "integrated waste management strategy" between 2019 and 2022. The tight regulatory and enforcement atmosphere in Rwanda, combined with higher material import costs due to its land-locked status, has encouraged the development of several industrial initiatives to recycle plastics.

It can therefore be seen that the Rwandan approach to plastic waste management is a mix of command-and-control policies and market-based instruments (Xie and Martin, 2022). The command-and-control elements comprise bans of single-use plastics and ethylene-based products, as well as standards regulating the manufacture of plastic products in-country. The market-based instruments include taxes, fees and subsidies. These provide incentives for green manufacturing and mobilisation of funds to run and maintain plastic waste management systems and infrastructures.

2.3 Uganda

While the latest data is not available, as of 2019, Uganda was reported to have imported 8,768,103 tonnes annually (Wandeka et al., 2022). A substantial portion of plastic wastes in Uganda is related to packaging. According to recent estimates, plastic packaging constitutes about 90% of all packaging in Uganda, and about 600 tonnes of plastic packaging is consumed daily (Wandeka et al., 2022). Unlike Rwanda, Uganda has not imposed an outright ban on single-use plastic. Instead, under the standard requirements published by the Ugandan National Bureau of Standards (UNBS), Uganda "prohibits the importation, export, local manufacture, use or reuse of categories of plastic carrier bags or plastic products made of polymers of polyethene or polypropylene below thirty (30) microns" (UNBS, 2021). Even in the absence of an outright ban, compliance is a significant challenge for the Ugandan authorities. The UNBS reported that following an inspection of 47 factories it undertook in 2021, 21 of them were found to be non-compliant and compelled to suspend production until they took corrective action (UNBS, 2021).

In addition to the partial ban described above, Uganda has also launched policy interventions to drive and monitor some forms of EPRs. Under the Ugandan requirements, plastic manufacturers are required to set up recycling plants and make arrangements for their plastic products to be returned for recycling (NEMA Uganda, 2020). At the international level, Uganda, in June 2021, joined the Clean Seas Campaign, a global movement of more than 60 countries committed to ending marine plastic pollution from source to sea (UNEP, 2021). Uganda is also working closely with the neighbouring countries of Tanzania and Kenya to tackle the growing menace of plastic pollution in the world's largest tropical lake, Lake Victoria, where microplastic is causing a huge havoc to marine ecosystem as they carry harmful chemicals and pollutants, in addition to direct threats on fish (The Flipflopi Project, 2022).

3 Methodology

In addition to secondary sources such as policy documents and reports, this chapter draws from qualitative primary data obtained through focus group discussions and semi-structured interviews of key stakeholders. A focus group discussion for five participants from technology startups across East Africa (Uganda, Kenya and Rwanda) was held in October 2020. This was held online using videoconferencing, recorded and transcribed. A briefing on the objectives of the DITCh plastic project preceded the focus group meetings to obtain relevant consent. Participants were then allowed to introduce themselves and their roles within the sector.

Following the completion of the focus groups, one participant was identified as an ideal candidate for further interviews as the insights they provided demonstrated their expertise and experience. Interviews were conducted online using videoconferencing, and all interviews were transcribed and recorded after receiving relevant consent from the participants. The transcripts of the focus group discussion and in-depth interviews were fed into NVivo 12 software where emerging insights and ideas were coded and thematically analysed.

4 Findings and discussion

The transcripts of the focus group and in-depth interviews highlight insights and perspectives on the following key themes: policy interventions and outcomes, challenges and opportunities for waste collection and private and informal sector contributions.

4.1 Policy interventions and outcomes

One of the key areas of interest in discussions about the progress of the campaign for a circular economy on the African continent is the importance of policy interventions and political will on the part of national governments to launch and implement necessary interventions. These interventions fall under two broad categories: prohibitions and incentives. Participants in the focus group reflected on the impact of single-use plastic ban in Rwanda:

> If we look at the family we start to get a grasp of how this problem is happening. But even in Rwanda, where we ban single use plastics, the invitation to pursue alternative packaging has actually been slow.
>
> *Focus Group, October 2020*

While Rwanda is often held up as an exemplar of successful government policies on environment and sustainability, the above comment from the focus group underlines the limitations of bans and prohibitions. Instead, policymakers need a carrots and sticks approach, where fines and bans are complemented with incentives and rewards for alternative production approaches and consumption habits. In line with this, another focus group participant highlighted the importance of government policies to drive market demands for circular products:

> There sometimes need to be some sort of push for the demand side, that's encouraged by the government. So one thing that came up is that if we could just have some legislation that requires a certain amount of recycled content in construction, for example, can make a huge difference.
>
> *Focus Group, October 2020*

As other studies have found, strategic public procurements and tax incentives can be used by governments to drive demand and encourage producers to use recyclates, rather than virgin materials, for the manufacture of plastic products (Hart et al., 2019). These "carrots" work better along with "sticks" like bans and fines. The use of incentives applies to producers, consumers and ordinary citizens. In this respect, digital innovations, such as blockchains, have been used in both developed and developing countries to mobilise and incentivise citizens to actively participate in the drive towards a circular economy (Ajwani-Ramchandani et al., 2021). In Spain, a virtual reward token was created to incentivise families to recycle and a webapp was created to enable them record recycled plastics (Gibovic and Bikfalvi, 2021). The need for policy interventions such as public procurement assumes greater strategic significance considering that recycling of certain polymers is not ordinarily profitable, even with high rates of plastic waste collection (Galati and Scalenghe, 2021).

4.2   Challenges and opportunities for waste collection

The respondents highlighted a wide range of logistical, practical and cultural challenges that are hampering efficiency of waste collection across their respective countries:

> I would say the biggest challenge is the culture. The culture of waste handling. We have companies that are doing waste collection, but still they do it unprofessionally, so that is a big challenge. We have no waste management professionals in Rwanda. That is a big challenge, I would say.
>
> *Rwanda Civil Society Focus Group, October 2020*

> I think it's for me mostly related to the waste separation. If you want to add value two ways, if you want to recycle the waste, we should separate them, yeah, that's my point.
>
> *Rwanda Civil Society Focus Group, October 2020*

> This (waste management) sector is really characterised by inefficiency and irregularities in waste collections. There is very low waste collection coverage and the other big problem is that there is a lack of household data (in Uganda). You know, there is some data out there, some statistics, but household data and which houses?
>
> *Uganda Focus Group, October 2020*

The feedback from the focus group participants reinforces the argument for a multi-sectoral, multi-stakeholder approach to sustainable plastic waste management. Top-down policies and regulations are not sufficient, in isolation. Public and private sector organisations need to work in dynamic synergy with the academia and non-governmental organisations (NGOs) to change culture and attitudes to plastic waste using a mix of public awareness campaigns, policies and innovations to change minds and redirect entrenched linear habits towards circularity. In order to address some of the key challenges highlighted above, a number of tech startups are stepping up with innovative ideas and products to tackle the challenges. This is exemplified by the initiatives and contributions of Yo Waste, a Kampala-based startup whose platforms and products are helping to connect households and businesses with other waste management players. They are doing this through three key platforms and products, as the founder summarises:

Yo-Waste Connect: For households and businesses to schedule waste pickup

Yo-Waste Hauler: For drivers and those who collect the waste. They enter data on the kind of waste collected and indicate when a job is completed. Yo-Waste plans to sell this data to governmental organisations or MNCs like MTN & Airtel.

Yo-Waste Cloud Platform: For bigger companies that have multiple pickup points and want to sign up as customers and for larger waste management companies who sign up to offer services. There is a dashboard for visualizing, managing and assigning job.

CEO, Yo Waste, Uganda, November 2020

Yo Waste's products exemplify the potentials of digital innovations in the circular plastic campaign. By linking up different stakeholders via digital platforms, innovators like Yo Waste are able to drive efficiency, reduce transaction costs and create new opportunities for waste collectors and recyclers (Oyinlola et al., 2022). In other words, digital innovations can invigorate the ecosystem for the circular plastic economy, thereby helping to realise the full benefits of government policy interventions (Kolade et al., 2022).

4.3   Private and informal sector contributions

Both focus group participants and interview respondents emphasised the importance of non-governmental actors, especially corporate actors, in the drive towards a circular economy. Equally important, there is a recognition for the role of informal actors, whose contributions are currently not optimally realised due in part to weak organisation and lack of empowerment:

We were engaging some international investors that come from the private side, but also some institutional investors who have large scale climate change or kind of funds to protect the environment and they had a discussion about the investment climate for these types of things and it was actually our international investor who highlighted this pointed out that.

Rwanda Academia Focus Group, October 2020

A lack of investable private projects, and sometimes this is complicated, complicated by the involvement of the informal sector being so important. So organising that informal sector seemed like a challenge that the investors were interested in.

Rwanda Academia Focus Group

And maybe then you also encourage the informal sector to collect more waste. And also it's a very important fact to understand that most of the people that are in the informal sector are just unemployed people and also very very poor usually …, they are unemployed they are poor and waste collection is maybe informal … I think maybe if you do it in this way that you also have maybe a return back scheme for maybe the bigger plastics, Maybe that could also benefit these informal sectors somehow.

Rwanda Civil Society Focus Group, October 2020

As the comments above show, public policy must have clear links with private and third sector contributions (Mugambe et al., 2022). Increasingly large corporations and manufacturers are giving greater attention to sustainability and circularity agenda. This is partly as a result of growing public awareness and scrutiny of large corporations about commitments to environmental and sustainability issues. The contributions of big corporations and plastic manufacturers should not be measured only in terms of outward-facing investments, because this effectively leaves the responsibility on other actors to clean up the mess brought about by linear and non-environment-friendly production practices. Instead, big corporations should also be scrutinised in terms of internal innovation, experimentation and adoption of circular business models in design, production and value delivery (Bocken et al., 2018). Plastic manufacturers need to rethink their value propositions and focus attention on using minimal resources for a maximum period of time in the process of delivering optimal value for end-users (Geissdoerfer et al., 2020).

Finally, as the focus group participants highlighted, the contributions of the informal sector cannot be understated in the drive towards the circular plastic economy. These otherwise invisible and unrecognised actors, who are typically driven to these roles through sheer necessity, are critical to successful transition to a circular economy through a wide range of activities including waste collection and recycling (Korsunova et al., 2022). With the right support and interventions, they offer a promising and effective pathway to an inclusive circular plastic economy, especially in low- and middle-income countries where waste collection and recycling facilities are limited. Interventions can be aimed at reducing barriers to waste collection, improving income opportunities for informal waste collectors and recyclers and increasing quality of materials (Velis et al., 2022). These empowerments will give them economic visibility and dignity, in order to maximise their potentials in the circular economy ecosystems.

5   Conclusion

This chapter highlights the critical importance of a multi-stakeholder approach, across a whole spectrum of the economy and society, to a circular plastic economy in the East African region. This chapter begins with a detailed discussion of the policy and contextual peculiarities of three East African countries: Kenya, Rwanda and Uganda. It describes the varying levels of policy success and the country-specific contexts that illuminates this. This chapter then presents primary qualitative data obtained from focus groups and in-depth interviews of participants across the East African region. This data highlights three important points: Firstly, targeted policymaking and political will make a significant difference in the drive towards a circular plastic economy, because these set the tone for other stakeholders in the private and third sectors. However, the results of policy interventions are mixed across countries. Rwanda appears to show the highest levels of policy success, but even the Rwandan government has had to grapple with entrenched cultural barriers and attitudinal obstacles to the circular economy. The success of policy interventions in countries like Uganda and Kenya is influenced by a range of geographical and political factors. Secondly, digital innovators are making significant impacts by using digital tools and platforms to mobilise and link key stakeholders and actors in the circular plastic ecosystem. Finally, the potentials of the private and informal sector actors are currently underutilised. With better organisations and the right incentives, informal sector operators can be better empowered to contribute to successful transition to a circular plastic economy across the East African region.

Acknowledgement

This work was supported by the United Kingdom Research and Innovation (UKRI) Global Challenges Researches Fund (GCRF) under Grant EP/T029846/1.

References

African Development Bank, 2021. East Africa Economic Outlook, Africa Economic Outlook Report. Africa Development Bank.

Ajwani-Ramchandani, R., Figueira, S., Torres de Oliveira, R., Jha, S., 2021. Enhancing the circular and modified linear economy: The importance of blockchain for developing economies. *Resour Conserv Recycl* 168, 105468. https://doi.org/10.1016/j.resconrec.2021.105468

Babayemi, J.O., Nnorom, I.C., Osibanjo, O., Weber, R., 2019. Ensuring sustainability in plastics use in Africa: Consumption, waste generation, and projections. *Environ Sci Eur* 31, 60. https://doi.org/10.1186/s12302-019-0254-5

Behuria, P., 2019. The comparative political economy of plastic bag bans in East Africa: Why implementation has varied in Rwanda, Kenya and Uganda. Global Development Institute Working Paper Series 1–30.

Bocken, N.M.P., Schuit, C.S.C., Kraaijenhagen, C., 2018. Experimenting with a circular business model: Lessons from eight cases. *Environ Innov Soc Transit* 28, 79–95. https://doi.org/10.1016/j.eist.2018.02.001

Galati, A., Scalenghe, R., 2021. Plastic end-of-life alternatives, with a focus on the agricultural sector. *Curr Opin Chem Eng.* https://doi.org/10.1016/j.coche.2021.100681

Geissdoerfer, M., Pieroni, M.P.P., Pigosso, D.C.A., Soufani, K., 2020. Circular business models: A review. *J Clean Prod* 277, 123741. https://doi.org/10.1016/j.jcle pro.2020.123741

Gibovic, D., Bikfalvi, A., 2021. Incentives for plastic recycling: How to engage citizens in active collection. Empirical evidence from Spain. *Recycling* 6, 29. https://doi.org/10.3390/recycling6020029

Griffin, M., Karasik, R., 2022. *Plastic Pollution Policy Country Profile: Kenya.* Nicholas Institute for Environmental Policy Solutions.

Hart, J., Adams, K., Giesekam, J., Tingley, D.D., Pomponi, F., 2019. Barriers and drivers in a circular economy: The case of the built environment. *Procedia CIRP* 80, 619–624. https://doi.org/10.1016/j.procir.2018.12.015

Kenyan Association of Manufacturers (KAM), 2019. Accelerating a circular economy in Kenya, Kenya plastic action plan. Available at: https://kam.co.ke/bfd_download/kenya-plastic-action-plan/ (Accessed: 23 March 2023).

Kolade, O., Odumuyiwa, V., Abolfathi, S., Schröder, P., Wakunuma, K., Akanmu, I., Whitehead, T., Tijani, B., Oyinlola, M., 2022. Technology acceptance and readiness of stakeholders for transitioning to a circular plastic economy in Africa. *Technol Forecast Soc Change* 183. https://doi.org/10.1016/j.techfore.2022.121954

Korsunova, A., Halme, M., Kourula, A., Levänen, J., Lima-Toivanen, M., 2022. Necessity-driven circular economy in low-income contexts: How informal sector practices retain value for circularity. *Global Environmental Change* 76. https://doi.org/10.1016/j.gloenv cha.2022.102573

Mugambe, R.K., Nuwematsiko, R., Ssekamatte, T., Nkurunziza, A.G., Wagaba, B., Isunju, J.B., Wafula, S.T., Nabaasa, H., Katongole, C.B., Atuyambe, L.M., Buregyeya, E., 2022. Drivers of solid waste segregation and recycling in Kampala Slums, Uganda: A qualitative exploration using the behavior centered design model. *Int J Environ Res Public Health* 19. https://doi.org/10.3390/ijerph191710947

NEMA Uganda, 2020. Uganda: Statement on the management of plastic pollution. p. 3. Available at: https://apps1.unep.org/resolution/uploads/uganda_statement.pdf (Accessed: 23 March 2023).

Oyake-Ombis, L., van Vliet, B.J.M., Mol, A.P.J., 2015. Managing plastic waste in East Africa: Niche innovations in plastic production and solid waste. *Habitat Int* 48, 188–197. https://doi.org/10.1016/j.habitatint.2015.03.019

Oyinlola, M., Schröder, P., Whitehead, T., Kolade, O., Wakunuma, K., Sharifi, S., Rawn, B., Odumuyiwa, V., Lendelvo, S., Brighty, G., Tijani, B., Jaiyeola, T., Lindunda, L., Mtonga, R., Abolfathi, S., 2022. Digital innovations for transitioning to circular plastic value chains in Africa. *Africa J Manage* 8, 83–108. https://doi.org/10.1080/23322 373.2021.1999750

Regional Economic Department Kenya, 2022. Waste management in East Africa. Nairobi Kenya. Available at: https://www.tresor.economie.gouv.fr/Articles/da8ae3d2-6292-46ca-b32d-f1b93a9c86ac/files/e2ee9b46-551b-4f08-8dd5-cc46f4ac42c9 (Accessed: 23 March 2023).

The Flipflopi Project, 2022. Microplastics found in 100% of locations in Lake Victoria: Results just in [WWW Document]. Blog. www.theflipflopi.com/blog/2022/6/13/microplastics-found-in-100-of-locations-in-lake-victoria-results-just-in (accessed 12.6.22).

UNBS, 2021. Enforcement of the ban on plastic carrier bags below 30 microns. Available at: https://www.nema.go.ug/sites/default/files/Press%20release-Kavera_NEMA_U NBS%20enforcement_2.pdf (Accessed: 23 March 2023).

UNEP, 2021. Uganda joins Clean Seas Campaign to keep plastic pollution out of its lakes and rivers [WWW Document]. Press Release. www.unep.org/news-and-stories/press-release/uganda-joins-clean-seas-campaign-keep-plastic-pollution-out-its   (accessed 12.6.22).

UNIDO, 2022. Study on plastic value chain in Kenya. Available at: https://www.unido.org/sites/default/files/files/2022-01/Plastic_value_chain_in_Kenya.pdf (Accessed: 23 March 2023).

Velis, C.A., Hardesty, B.D., Cottom, J.W., Wilcox, C., 2022. Enabling the informal recycling sector to prevent plastic pollution and deliver an inclusive circular economy. *Environ Sci Policy* 138, 20–25. https://doi.org/10.1016/j.envsci.2022.09.008

Wandeka, C.M., Kiggundu, N., Mutumba, R., 2022. Plastic packaging: A study on plastic imports in Uganda. *Int J Sci Adv* 3, 19–26. https://doi.org/10.51542/ijscia.v3i1.2

Xie, J., Martin, J., 2022. *Plastic Waste Management in Rwanda: An Ex-post Policy Analysis.* Washington, DC: World Bank. https://openknowledge.worldbank.org/entities/publication/2f51d4ab-064e-55f5-90b8-92eafff608af License: CC BY 3.0 IGO

10

THE APPLICATION OF DIGITAL TECHNOLOGY IN CIRCULAR PLASTIC ECONOMY IN SOUTHERN AFRICA

Case studies of waste management start-ups from Namibia and Zambia

Selma Lendelvo, Mecthilde Pinto, Florensa Amadhila, Luzé Kloppers-Mouton, Chifungu Samazaka, Raili Hasheela and John Sifani

1 Introduction

The management of solid waste continues to be a global challenge, and in the Global South (developing countries), the challenge is anticipated to be of a greater magnitude in the coming years (Mwanza & Mbohwa, 2019). Several authors have documented the wide-ranging challenges that underpin plastic waste management. These include, but are not limited to, a lack of financial resources allocated to the waste management sector, lack of recycling systems and facilities and limited knowledge on waste management (Arbulú et al., 2016; Gobbi et al., 2017; Elsaid & Aghezzaf, 2015; Reinhart et al., 2016).

Van Niekerk and Weghmann (2019) contend that many African countries frequently practice open dumping and burning as a form of waste disposal, with almost half of the estimated waste generated ending up on controlled and uncontrolled open dumpsites (van Niekerk & Weghmann, 2019). It is reported that from these 50 biggest dumpsites globally, Africa contributes about 19 dumpsites, mostly located in sub-Saharan Africa (African Union Development Agency, 2021). Further, it is estimated that only about 30% of the waste generated in Africa is disposed of in formal landfill sites (Silpa et al., 2018). This posits that large quantities of waste material generated in Africa remain unmanaged; uncollected; and accumulating within towns, cities and villages. For example, packaging, which is primarily plastic materials, is among the major waste in Southern African countries, where South Africa alone represents 55% of plastic waste in the region (Sadan & De Kock, 2021).

DOI: 10.4324/9781003278443-12

These open dumping and burning practices pose major environmental pollution and threaten the health and the well-being of the communities (Sadan & De Kock, 2021; Silpa et al., 2018). What makes it worse is the fact that these dumping are done informally as the waste collection system is inadequate. Accordingly, if these African countries do not find sustainable waste disposal and management systems, negative environmental impacts such as environmental degradation and reduced aesthetic views may be inevitable (Sadan & De Kock, 2021). Further, these wastes may contaminate water systems posing health risks to humans and animals ultimately endangering land and aquatic life (Silpa et al., 2018). Therefore, it is opined that human exposure to poorly managed waste sites contributes to serious health risks across the African continent (African Union Development Agency, 2021; Sadan & De Kock, 2021).

Compared to the other types of waste, the impacts of plastic waste are the worst because they are non-biodegradable, and it takes many years to decompose (Pinto da Costa et al., 2020). Regrettably, even when broken into smaller pieces, they still release toxins into the environment, whether in the soil or in the water. When burned, they release toxic substances that cause ambient air pollution (Sadan & De Kock, 2021). This type of impact can be cumulative and result in diseases, greenhouse gas (GHG) emissions or underground water pollution. In the marine environment, plastic litter can negatively impact tourism, as they cause environmental pollution (Alabi et al., 2019). The national economy can be negatively impacted because of plastic litter through the reduction of tourists. Given the above challenges associated with waste, significant efforts must be placed on strategies and technologies that can potentially reduce the consequences thereof (Adeniran et al., 2022).

Several actors are involved in waste management; however, the municipality plays a significant role especially within local authorities. In addition, with the increasing migrations from rural to urban areas, many towns have expanded, exacerbating the high informal settlement dwellers and their waste activities. This hampers the municipal and institutional capacities to deal with the waste management services in these areas (African Union Development Agency, 2021).

Subsequently, alternative approaches have been and are being adopted to address the waste management challenge. For example, in Southern Africa, countries like Zambia have developed policies and strategies to recover and improve recycling and waste management in general (Mwanza et al., 2018). In the fight against waste, African countries have gone as far as involving private companies, small and medium enterprises (SMEs), independent operating start-ups or outsourced companies by the municipality, cooperatives of informal workers, non-governmental organisations (NGOs), community groups and individual waste pickers (van Niekerk & Weghmann, 2019). These actors handle waste in the different phases or a combination thereof, inclusive of collection, processing (e.g., recycling) and finally, disposal to reduce the accumulation of waste quantities.

Conservatively, waste reduction used local/traditional knowledge and a trash collection approach. However, the use of traditional local approaches is slowly being overtaken by the application of modern digital technological innovation such as the use of mobile apps, mobile payment systems, artificial intelligences in solving humanity's daily challenges, teaching and learning, creating awareness and advocacy (Mundia et al., 2021; Wilson et al., 2021). Additionally, such technologies are also being used for networking and community education, while at the same time contributing to increased waste management efforts, such as reuse, recycling and recovery, especially for plastics (Babayemi et al., 2019; Godfrey et al., 2018; Xevgenos et al., 2015).

Most of these technologies are still in the testing phases; however, they are anticipated to play an essential role in waste management through circular economy (CE) approaches and tackling waste (United Nations Development Programme, 2019). Today, such interventions are emergent across Africa and utilised to decrease unsustainable waste disposal where insufficient waste collection coverage is eminent (van Niekerk & Weghmann, 2019). However, due to the knowledge gap, the adoption of digital technologies for waste management is still relatively new, particularly in Southern Africa. A majority of the stakeholders including municipalities ICT developers and citizens still require an understanding of digital innovation usability and application (Ringenson et al., 2018). Feijao et al. (2021) noted that while digital innovations are expanding rapidly, the demand for digital skills have become high, however supply of digital skills is low, creating a 'digital skill gap'. Resulting in difficulties adopting and implementing new digital innovations in developing countries, enabling different business functions to adopt digital innovations that are mainly concentrated on manual and semi-automated processes as compared to adopting fully-automated and ICT-enabled, or digitally-enabled technological innovations (Avenyo et al. 2022). Additionally, Avenyo et al. (2022) show that labour-force related constraints such as lack of human computer interaction skill reduces the likelihood to adopt digitalization, while skilled human capital particularly science, technology, engineering and math (STEM) qualifications have a higher likelihood to adopt digital innovations. Suggesting the importance of knowledge generation and STEM skills in the labour force (Avenyo et al. 2022). Therefore, this chapter asserts that this is also significant in the drive towards CPE in fostering the adoption of new digital innovations. Further, Mundia et al. (2021) bring the dynamism and look at bottlenecks, impeding smooth acceleration of grassroots innovation in Namibia, among other aspects including technological savvy.

Although digital innovations in waste management can be defined in various ways, this chapter views this kind of innovation as the use of new technologies, apps, smartphones and social media by citizens to communicate waste management issues through capturing, reporting and tracking waste in various areas or communities (Adeniran et al., 2022). Remarkably, several African countries are progressively leveraging waste recycling technologies; these are predominantly evident among SME start-ups which aim to actively manage waste through CE approaches (Oyinlola et al., 2022). Digitising waste management in Southern Africa among start-ups has dual benefits both for the environment and active participation of citizens, which has the ability to accelerate waste reduction among communities and municipal areas (African Union Development Agency, 2021).

The Southern African region is among the less population densities in Africa, but also heavily relying on the plastic-based market (Adeniran et al., 2022). In addition, this region is characterised by young economies with emerging SME start-ups. This chapter presents the results of the assessment made on the waste management efforts being practised by three different waste management start-ups, two from Namibia and one from Zambia. This chapter has paid particular attention to the application of digital technology in waste management by the three start-ups and their contribution to CE.

The objectives of the assessment were to:

(I) Identify the current waste management efforts by the local waste management start-ups;
(II) Identify the different challenges experienced by the three start-ups in the face of circular plastic economy (CPE);
(III) Assess the achievements and drivers for new technological innovations for improving plastic waste management; and
(IV) Recommend the way forward regarding waste management using digital innovation.

2 Methods

The analytical framework adopted for this study is the comparative case study approach. Comparative case studies cover two or more cases in a way that produces more generalisable knowledge (Goodrick, 2014). This analytical framework documents interventions that address waste management by applying digital innovations in Southern Africa. This study adopted this approach to produce knowledge that can be used to generalise questions and determine relevant lessons from the existing start-ups that aim to address waste management and pollution in Southern Africa. We selected three case studies for an intensive and in-depth qualitative analysis based on the scope of DITCh plastic project. These three start-ups were identified based on their (1) geographical commonalities being Southern Africa. (2) due convenience and availability of information, and access to interview partners as the start-ups formed part of the stakeholders available for information sharing and capacity building sessions hosted by of DITCh plastic project coordinators. The first step was opportunity mapping of the existing innovation initiatives and entrepreneurial models that apply digital technologies to address plastic waste. Thematic areas for further assessment were conducted on the key aspects, which are displayed in table and graphic format.

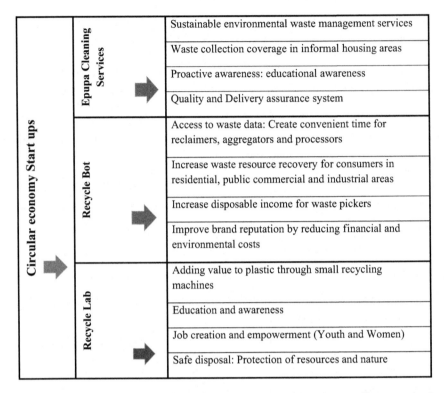

FIGURE 10.1 The focal areas of the three start-ups seeking to contribute to plastic economy and their functions

Secondly, a qualitative method of data collection and analysis was used for the assessment. Data was drawn from key informant interviews (KIIs) conducted with the founders of the three start-ups, namely Epupa Cleaning Services and The Recycling Lab from Namibia and Recyclebot from Zambia, during the month of October 2020. Due to the COVID-19 pandemic regulations, data was digitally collected via Zoom and through emails. The collected data was analysed through group themes, using Microsoft Excel. Direct quotations/narratives of the start-up founders are also documented.

3 Results and discussion

This section presents the findings regarding the opportunistic areas maximised by the start-ups for transitioning from linear to CE in Africa (Figure 10.1).

4 Case studies

Opportunities presented by digital innovations have encouraged the development of start-ups that employ these innovations for waste management across the globe including Southern Africa. This section presents three start-ups that are making major investments in digital innovation to promote waste management: (1) Epupa Cleaning Services in Namibia, a start-up that primarily focuses on delivering waste management services to the public, while promoting alternative long-term solutions for plastic waste; (2) The Recycling Lab cc in Namibia, a technology support-focused start-up, that seeks to facilitate the methods of safe disposal of waste, while promoting nature preservation through the use of small recycling machines; and (3) Recyclebot in Zambia, an online app material recovery platform that aims to offer waste management pickers convenience by reducing the time spent aggregating the recovered waste by 80%.

4.1 Epupa Cleaning Services cc

Epupa Cleaning Services is a Namibian-owned CE start-up that primarily focuses on delivering waste management services to the public, while promoting alternative long-term solutions for plastic waste. Since its establishment in 2001, Epupa has employed about 180 people and trained 60 people, inclusive of men, women and the youth to be independent entrepreneurs in waste management services such as cleaning, street sweeping and refuse removal in the suburbs of City of Windhoek.

Epupa offers a wide range of waste management-related services, like recycling (e.g., plastics, paper, glass and cans), hospital ward or street cleaning and sweeping, removal of biohazardous chemical waste and organic and medical waste disposal. They supply 240-litre wheelie bins and skip containers (6 and 9 m³) and offer the removal of skips. They also level buildings dispose of the rubble, and they do garden refuse disposal to various sites around Windhoek. Epupa contributes to the reduction of plastic waste through the promotional use of paper or cloth bags and the use of sustainably recycled plastic waste. The company is actively involved in community education and awareness interventions seeking to motivate community members to become active partakers of environmental management and adaptors of eco-friendly practices. Moreover, the company has plans to ensure that plastic waste is recycled into reusable products, such as 240-litre wheelie bins, plastic chairs, pole refuse bins, interlocks, etc. Thus, the company is looking to invest in manufacturing.

Epupa has identified the need to establish a waste storage centre "recycle bank" where waste collected can be stored for processing; and Epupa will transact funds to the waste collectors for the waste collected. In terms of digital innovation, Epupa has identified key strategies for investing in digital technology to design an app for informal small traders of household and commercial waste, which they can use to upload the quantity of recycled waste collected in a specific month. Mobile apps such as WhatApp are being mainly used in this start-up to communicate with the workers, informal traders and commercial traders. However, this technology presents shortcomings as it is not technologically assembled to weigh waste, transact funds and capture data of waste collected by informal traders. Therefore, Epupa has identified the need to develop and implement digital innovations such as an online app which offers informal small traders the opportunity to capture their waste collected, weigh it and earn points which can be translated into economic gains.

4.2 The Recycling Lab cc

The Recycling Lab cc is a recently established company that seeks to facilitate the methods of safe disposal of waste, while promoting nature preservation. It is dedicated to adding value to plastic waste which it achieves through its technological support, through the use of small recycling machines, network building for the CE and community education. Furthermore, The Recycling Lab cc plans to establish waste separation stations, where the unemployed youth in rural areas can dispose their collected recyclable waste and be sorted accordingly. This start-up has identified key strategies for technological investment such as shredding machines, recycling on site or making plastic pellets from waste, which can be sold back to manufacturers.

Most importantly, this start-up promotes the use of digital tools for measuring and quantifying accurate data to inform decision-making processes in the country and to contribute to the development of legislations and policies. The Recycling Lab cc has created an online data platform, aimed at recording various data that is collected from all the product material types that come into Namibia from importing companies and to collect data from local manufacturers and traders that import their raw materials. This initiative of collection and uploading of waste data from commercial traders was indicated to be challenging due to the lengthy administrative process to acquire data and companies protecting the confidentiality of their data sets.

4.3 Recyclebot

Recyclebot is a start-up that has been established in Zambia, which has developed an online app material recovery platform that aims to offer waste management pickers convenience by reducing the time spent aggregating the recovered waste by 80%. The app can be used to sell and to crowdsource the valued waste, with little effort and great reliability.

Since the development of the app, about 239 people in Zambia, 86 people in South Africa and 6 in Nigeria have utilised Recyclebot to date. Recyclebot has employed about 16 people: 3 working full-time and 13 working part-time (20–30% level of efforts); of which 4 are female. Due to ease of access of the online app which is also embedded in android and iOS apps, about 331 local communities and 5 commercial companies have utilised the app to transact their waste. Access to the website and web app requires internet access. The majority of reclaimers/waste pickers do not have smartphones or cannot afford the data fee, and they usually share smartphones to access the web or use the SMS Waste Aggregator which allows users to trade waste offline, without the need for smartphones or internet access.

The key strategy for investment and expansion includes Waste Marketplace which is an online marketplace that connects waste pickers, aggregators and manufacturers for efficient and fair trade of recovered waste. This web and mobile app gives users a customised waste storefront and a chance to give support to buyers; manage products, orders and payments; view store sales reports in detail; generate statements; and get a complete overview of their store's performance from the forefront. Additionally, Recyclebot has identified the need for Waste Voice Assistant and CHATBOT, which is a behavioural change app which allows users to measure, understand and reduce their waste by equipping them with daily, micro-learning about CE. This virtual assistant speaks multiple languages and will support consumers and waste workers from start to finish, in their material recovery processes, using a unique combination of both personalised practical teachings in addition to mindset management. Skills across micro-manufacturing, marketing, accounting, basic HR, strategy and communication will be taught using artificial emotional intelligence, which are deployed across easy-to-access social-media platforms (such as Facebook, WhatsApp, etc.). To further contribute to establishing relations between private waste collection operators and recyclers, Recyclebot has plans to establish Waste Material Tracker, a logistics app which will allow recyclers to assign transporters to materials and track package deliveries with geo-fencing and time-tracking. The founders for this app have outlined the key drivers for new technological innovations for improving plastic waste management and achievements for transitioning to the CE for plastics.

5 Discussions

5.1 Contribution of digital waste management to the sustainable development agenda

The assessment highlighted six opportunity areas / key drivers for improving waste management and achievements for transitioning to a CPE. The identified thematic or opportunity areas are (i) environmental sustainability – this reflects the importance of transition into innovative waste management practices to future environmental integrity; (ii) technological and digital innovations – the start-ups integrated contemporary technology and digital innovations for public or citizen inclusions and acceleration waste solution in communities; (iii) economic significance – the analysis also indicated the economic value of waste and how the involvement of community members can transform their lives and surroundings; (iv) employment creation and enterprise development – waste management enterprises were able to employ people or subcontract other SMEs; (v) livelihood improvement – jobs and income received from waste management efforts had

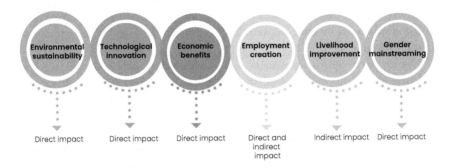

Direct impact   Direct impact   Direct impact   Direct and indirect impact   Indirect impact   Direct impact

FIGURE 10.2   CPE start–up's direct and indirect impacts

visible livelihood improvement on households; and (vi) gender equality – despite being a male activity, in the case studies of this chapter, the female-owned start-ups were a clear indication of women involvement in the sector (Figure 10.2).

This chapter shows that if digital innovations are not meaningful to the general society, who are the ultimate beneficiaries, then the impacts of such innovations are of no effect. In view of environmental sustainability, there is no standard definition of the concept; however, it refers to the relationship between human beings and the ecosystem they inhabit, while ensuring provision of clean air, clean water and a clean and productive land (Morelli, 2011). For the environment to be sustainable, it requires an intervention of the society that through significant efforts in environmental management can lead to sustainable environmental resources, which the economy depends on. As an ultimate result, livelihoods can improve, while at the same time employment opportunities can be created. Moreover, technological innovations can contribute to solving environmental, social and economic problems which cannot be solved through human direct efforts. Particularly in terms of environmental management, technological innovation can increase economic competitiveness, as a result of economic implementing activities aiming to contribute to environmental protection through technological means (Diaconu, 2011). Furthermore, ensuring environmental sustainability, technological innovation, economic benefits, employment creation and livelihood improvement may require gender mainstreaming, which entails an integration of the gender perspective towards achieving gender equality (EIGE, 2016).

Looking at the management of plastic waste in Southern Africa, circular plastic economic growth is possible, particularly through digital innovation. Based on the interviews with the three start-up founders, the current CPE opportunities have demonstrated high potential to contribute to environmental sustainability. They all agreed that environmental sustainability is the mandate of their business operations, while pursuing the vision to contribute to the long-term sustainable development goals (SDGs) and to Namibia's Vision 2030 (UN, 2020a). It is worth noting that waste management is one of the critical aspects of environmental management, and addressing it will undoubtedly contribute to the achievement of the Sustainable Development Agenda, under which the 17 goals have been set. Therefore, as an effort to achieve environmental sustainability, it should seek to contribute to the following SDGs: 6 (clean water and sanitation), 7 (affordable and clean energy), 9 (industry, innovation and infrastructure), 12 (responsible consumption and production), 13 (climate action), 14 (life below water) and 15 (life on land) (UN, 2020a). While environmental sustainability will contribute to the seven goals, digital innovation will contribute to SDG 9.

All three case studies emphasised the need for communities, both the informal and formal waste managers, to gain economic gains from waste management practices such as recycling. This can be achieved through the collection of recyclable materials in their respective communities and disposal sites; once collected, it can be sold to recycling companies such as the three start-ups to generate income. The long-term impacts of economic gains thus contribute to SDG 8, decent work and economic growth, the goal that is promoting inclusive and sustainable economic growth, full of productive employment and decent work for all (UN, 2020a).

Currently, there are several small-scale traders and companies that are recycling and processing waste, to create employment opportunities. They have teams that go around collecting waste, which they sort and do a semi-process for recycling. The creation of these job opportunities assists those who do not have formal qualifications. This chapter has suggested that while the transition towards CE in Africa is being built upon the lessons learnt and principles developed in northern countries, there are distinct differences in design, strategy and implementation of the CPE in Africa.

5.2 Constraints and challenges to adopting more circular approaches among start-ups

Access to funding was identified as a major constraint to transitioning to a CPE for these start-ups, existing parallel at the national and international levels. The lack of an enabling environment, for financial and other incentive-based streams, hinders entrepreneurs from setting up businesses (Desmond & Asamba, 2019). For instance, the Epupa start-up was financed through a commercial bank loan. Post establishment of businesses, progressive financing for waste services is a major factor shaping the quality of waste management services. For start-ups to deliver quality services, finances are required for maintenance and upkeep of equipment and technology.

The CE is a highly industrialised sector; hence, access to technological innovation and machinery is a key aspect for businesses to perform tasks related to plastic management such as recovery and recycling, product life extension, sharing platform and product and service (Desmond & Asamba, 2019). Technological innovation was pointed out to be a constraint, with views of the start-up founders being that there are high expectations and dependency by some governments on western or international technological advancement, to solve the waste issues in Africa. This has resulted in lack of technical ability and low-tech approach to waste management; as a result, locally feasible technological advances are not widely disseminated and with emerging governmental support have been slow or absent. Thus, in the promotion of inclusive waste management practices, digital innovations designed should be easy to use and locally acceptable, and available technology should be prioritised.

For instance, the founder of Recyclebot narrated:

> Typically, we need support on more technical infrastructure to be able to involve more youth/women led organizations in waste management, we also need technical assistance to roll-out flexible device financing options, R&D assistance to continue improving existing products and building new products to ensure we can deliver on our brand promise and compete against other players that are also moving to online models, etc.
>
> *Pers. Comm 26.07.2021*

The adoption of external technology, without adequate technical capability, can result in incompatible usage, leading to high cost as technology becomes un-scalable requiring high-tech skills and knowledge to be operated. This constraint further exacerbates the already existing deficit in social skills and aptitude at the national and community levels. This is because, the majority of workforce involved in waste management is from the informal sector, who are not suitably skilled and not equipped with appropriate technological know-how.

The founder of The Recycling Lab cc indicated:

> We strive to facilitate meaningful engagements with communities and destinations, but this comes with its own challenges. And to achieve this is a challenge in itself as ongoing education and awareness is vital. It is not a once off conversation, but an ongoing dialogue. The whole way of thinking about waste should be changed and this can't be done overnight.
>
> *Pers. Comm 26.07.2021*

Hence, leveraging locally feasible technology and implementing circular policies and business principles according to each in country-specific context or needs could give directions in building policies and improving each country's productivity (Desmond & Asamba, 2019; McKinsey Global Institute, 2019). Additionally, empirical studies are increasingly showing that a well-developed digital infrastructure characterised by technological accessibility has the potential to increase productivity by lowering transaction costs, reducing distance and time impacts while improving usage by integrating markets with global value chains through the use of information flows (McKinsey Global Institute, 2019).

Another major challenge stressed was the insufficient waste data and lack of access to existing data on waste. The start-up founders indicated that a data-driven waste management model, which incorporates autonomous sorting, e-commerce and automated logistics, is new to the sub-Saharan markets. This is mainly stimulated by the lack of co-ordinated efforts to implement the National Solid Waste Management Strategy, and lack of technology-enabled waste management system as the traditional usage of formal waste collection system was centralised towards waste collection, recycling and disposal with limited consideration for collection and storing data.

Collecting and storing of data related to waste was acknowledged to be an enabling opportunity that could contribute to sustainable waste management and transition to a CPE, as narrated below:

> Having access to this data, can enable us to make better and more informed decisions to create a greener future for all. Without data we can't see the whole picture. Without data, it is not possible to put in place thorough legislations and policies to help create a circular economy and better solutions to manage plastic waste.
>
> *Pers. Comm, The Recycling Lab founder, 27.07.21*

> A data-driven approach ensures that viable materials are recovered from multiple waste streams and across different geographical areas, and that the waste management service providers have in-depth economic value chain analysis, including market (both domestic and export) assessments, competition/profitability analyses, gender analyses, and strategies to improve competitiveness in light of market opportunities and constraints.
>
> *Pers. Comm, Recyclebot, founder, 27.07.21*

> Only minimal data and information are available on waste quantities and practices. Improved data are important to facilitate better planning and to monitor that improvements are implemented.
>
> *Pers. Comm, Epupa Cleaning Services, founder, 27.07.21*

Similarly, a study by Babayemi et al. (2018) reported that the connection between trade and inventory data on waste can eventually be used as a tool to develop counter-measures, improve prevention and management programmes and calculate recycling quotas. Figure 10.3 shows the identified CPE challenges experienced by entrepreneurs/Innovators.

There is no doubt that investing in waste management contributes to livelihood improvement. While the waste management initiatives are creating employment opportunities, they are contributing to many people's livelihoods. Most waste workers who are employed in waste management companies earn income through the employment opportunities offered to them. As a result, their livelihoods are

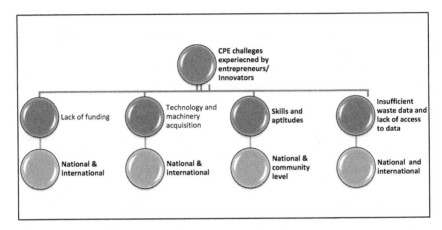

FIGURE 10.3   The identified CPE challenges experienced by entrepreneurs/innovators during the analysis of the three start-ups

improved, as they are able to depend on their income for survival (Rahman et al., 2017). Contribution of waste management to livelihoods also means promoting the attainment of SDG 10: reduce inequality, which is promoting the reduction of inequality within and between countries; and SDG 11: sustainable cities and communities, which seeks to make cities and human settlements inclusive, safe, resilient and sustainable (UN, 2020a).

The SDG agenda has set the period 2020–2030 as a decade of action, where all goals should be taken seriously, making sure that no one is left behind. This means, even with waste management, no specific type of gender is expected to participate, as it is a non-gender-specific effort. This contributes to SDG 5: gender equality, which is promoting the achievement of gender equality and the empowerment of all women and girls (UN, 2020b). However, innovation and technology seldom benefit women and men equally. This gender gap constrains efforts to achieve gender equality and women's empowerment and prevents women from becoming both developers and consumers of technology that addresses their needs (Van den Berg et al., 2020). To contribute towards gender equality, Canada requires departments and agencies to determine whether there is a potential gender issue in any policy, programme, initiative or service they propose before implementation, such methods would contribute towards reducing gender disparities, particularly in the design of technological and digital solutions (Van den Berg et al., 2020).

6   Conclusion and recommendation

The case studies in this chapter clearly illustrate that digitisation and technological transfer in waste management can have several benefits for enterprise development, environmental integrity and societal transformation. The three start-ups have demonstrated the applications of digital innovation and promotion of complementary waste management solutions within their respective countries and the importance of the promotion of a CPE locally. For instance, Recyclebot operates an online app material recovery platform, making waste pickers efficient by reducing the time spent aggregating the recovered waste, selling and crowdsourcing valued waste. Digital innovations for the CPE involve start-ups adding value to plastic waste through technological support like small recycling machines, building strong networks for CE, promoting the use of mobile phones and educating the community.

The analysis of the three start-ups also identified six opportunity areas for improving and digitising waste management in Southern Africa. These opportunity areas suggest the significant all-encompassing way of addressing aspects related to waste management including environmental sustainability, economic significance, employment creation, livelihood improvement and gender mainstreaming or empowerment and technology innovations. To achieve a CPE, the promotion of inclusive waste management practices and an easy-to-use/locally acceptable and available technology should be prioritised. This chapter shows that if digital innovations in waste management are to be meaningful to the general society and the replacing traditional plastic waste management practices, then the impacts of such innovations and the associated skills and knowledge should be made visible and shared at all levels. This study has contributed to knowledge creation on the CPE in Africa, particularly Southern Africa, and fills a gap identified by Desmond and Asamba (2019) that African case studies stay "hidden" as they have yet to be documented through academic research.

Acknowledgement

This work was supported by the United Kingdom Research and Innovation (UKRI) Global Challenges Research Fund (GCRF) under Grant EP/T029846/1.

References

Adeniran, A. A., Shakantu, W., & Ayesu-Koranteng, E. (2022). A proposed digital control system using a mobile application for municipal solid waste management in South Africa. *Waste Technology*, 10(1), 30–42. http://dx.doi.org/10.14710/wastech.10.1.30-42

African Union Development Agency (AUDA-NEPAD). (2021). What a waste: Innovations in Africa's waste material management. Retrieved on 20 November 2022, from www.nepad.org/blog/what-waste-innovations-africas-waste-material-management

Alabi, O. A., Ologbonjaye, K. I., Awosulu, O., & Alalade, O. (2019). Public and environmental health effects of plastic wastes disposal. *Journal of Toxicology and Risk Assessment*, 5, 021.

Arbulú, I., Lozano, J., & Rey-Maquieira, J. (2016). The challenges of municipal solid waste management systems provided by public-private partnerships in mature tourist destinations: The case of Mallorca. *Waste Management*, 51, 252–258.

Avenyo, E. K., Bell, J. F., & Nyamwena, J. (2022). Determinants of the Adoption of Digital Technologies in South African Manufacturing: Evidence from a Firm-level Survey. SARChI Industrial Development Working Paper Series WP 2022-13 ISBN 978-0-6398362-7-0

Babayemi, J. O., Nnorom, I. C., Osibanjo, O., & Weber, R. (2019). Ensuring sustainability in plastics use in Africa: Consumption, waste generation, and projections. *Environmental Sciences Europe*, 31, 60.

Babayemi, J. O., Ogundiran, M. B., Weber, R., & Osibanjo, O. (2018). Initial inventory of plastics imports in Nigeria as a basis for more sustainable management policies. *Journal of Health and Pollution*, 8(18), 6–20.

Desmond, P., & Asamba, M. (2019). Accelerating the transition to a circular economy in Africa: Case studies from Kenya and South Africa. In *The Circular Economy and the Global South* (pp. 152–172). Routledge.

Diaconu, M. (2011). *Technological Innovation: Concept, Process, Typology and Implications in the Economy*. Gheorghe Asachi Technical University of Isai. www.researchgate.net/publication/227364059

EIGE. (2016). *What Is Gender Equality?* European Institute of Gender Equality.

Elsaid, S., & Aghezzaf, E. H. (2015). A framework for sustainable waste management: challenges and opportunities. *Management Research Review*, 38(10), 1086–1097. DOI: 10.1108/MRR-11-2014-0264

Feijao, C., Flanagan, I., van Stolk, C., & Gunashekar, S. (2021). *The global digital skills gap: Current trends and future directions*. RAND Corporation. Retrieved from https://www.rand.org/pubs/research_reports/RRA1533-1.html

Gobbi, C. N., Sanches, V. M. L., Pacheco, E. B. A. V., Guimarães, M. J. O. C., & de Freitas, M. A. V. (2017). Management of plastic wastes at Brazilian ports and diagnosis of their generation, Marine. *Marine Pollution Bulletin*, 124, 67–73.

Godfrey, L., Nahman, A., Yonli, A. H., Gebremedhin, F. G., Katima, J. H., Gebremedhin, K. G., … & Richter, U. H. (2018). Africa waste management outlook. https://wedocs.unep.org/bitstream/handle/20.500.11822/25515/Africa_WMO_Summary.pdf?sequence=1&isAllowed=y

Goodrick, D. (2014). Comparative case studies, UNICEF. Retrieved from: http://devinfolive.info/impact_evaluation/img/downloads/Comparative_Case_Studies_ENG.pdf

McKinsey Global Institute. The great transformer: The impact of the internet on economic growth and prosperity. Available online: http://dese.ade.arkansas.gov/public/userfiles/Legislative_Services/Quality%20Digital%20Learning%20Study/Facts/McKinsey_Global_Institute-Impact_of_Internet_on_economic_ growth.pdf (accessed on 1 August 2019).

Morelli, J. (2011). Environmental sustainability: A definition for environmental professionals. *Journal of Environmental Sustainability*, 1(1), Article 2. DOI:10.14448/jes.01.0002. Available online: http://scholarworks.rit.edu/jes/vol1/iss1/2

Mundia, L., Sifani, J., Lupaphla, N., & Haipinge, J. (2021). Exploring bottlenecks towards accelerating grassroots Innovation in Namibia. *Namibia Journal for Research, Science and Technology*, 3, 9–13.

Mwanza, B. G., & Mbohwa, C. (2019). Technology and plastic recycling: Where are we in Zambia, Africa? International Association for Management of Technology in Managing Technology for Sustainable and Inclusive Growth Conference Proceedings.

Mwanza, B. G., Mbohwa, C., & Telukdarie, A. (2018). Strategies for the recovery and recycling of plastic solid waste (PSW): A focus on plastic manufacturing companies. *Procedia Manufacturing*, 21, 686–693.

Oyinlola, M., Schröder, P., Whitehead, T., Kolade, O., Wakunuma, K., Sharifi, S., ... & Abolfathi, S. (2022). Digital innovations for transitioning to circular plastic value chains in Africa. *Africa Journal of Management*, 8(1), 83–108.

Pinto da Costa, J., Rocha-Santos, T., & Duarte, P. A. C. (2020). *The environmental impacts of plastics and micro-plastics use, waste and pollution: EU and national measures.* Policy Department for Citizens' Rights and Constitutional Affairs Directorate-General for Internal Policies PE 658.279.

Rahman, M. Z., Siwar, C., & Begum, R. A. (2017). Achieving sustainable livelihood through solid waste management in Dhaka City. *International Journal of GEOMATE*, 12(30), 19–27. Special Issue on Science, Engineering & Environment, ISSN: 2186-2990, Japan.

Reinhart, D, Bollard, C. S., & Berge, N. (2016). Grand challenges – Management of municipal solid waste. *Waste Management*, 49, 1–2.

Ringenson, T., Höjer, M., Kramers, A., & Viggedal, A., (2018). Digitalisation and environmental aims in municipalities. *Sustainability*, 10(4), 1278. https://doi.org/10.3390/su10041278

Sadan, Z., & De Kock, L. (2021). *Plastic Pollution in Africa: Identifying Policy Gaps and Opportunities.* WWF South Africa, Cape Town, South Africa.

Silpa, K., Yao, L., Bhada-Tata, P., & Woerden, F. (2018). "What a Waste 2.0: A Global Snapshot of Solid Waste Management to 2050." Overview booklet. World Bank, Washington, DC. License: Creative Commons Attribution CC BY 3.0 IGO.

United Nations (UN) (2020a). The sustainable development goals report. United Nations. Retrieved from https://desapublications.un.org/publications/sustainable-developm ent-goals report-2022

United Nations (UN) (2020b). The 17 goals. Retrieved from https://sdgs.un.org/goals

United Nations Development Programme (2019). Annual report 2019. Retrieved from https://annualreport.undp.org/2019

Van den Berg, A. C., Giest, S. N., Groeneveld, S. M., & Kraaij, W. (2020). Inclusivity in online platforms: Recruitment strategies for improving participation of diverse sociodemographic groups. *Public Administration Review*, 80(6), 989–100.

van Niekerk, S., & Weghmann, V. (2019). Municipal solid waste management services in Africa. *Ferney-Voltaire: PSI (PSIRU Working Paper).* www.world-psi.org/en/municipal-solid-waste-management-servicesafrica

Wilson, M., Kitson, N., Beavor, A., Ali, M., Palfreman, J., & Makarem, N. (2021). *Digital Dividends in Plastic Recycling.* Global System for Mobile Communications Association.

Xevgenos, D., Papadaskalopoulou, C., Panaretou, V., Moustakas, K., & Malam, D. (2015). Success stories for recycling of MSW at municipal level. *Waste Biomass Valor*, 6, 657–684.

PART III

A Digitally Enabled Circular Plastic Economy

11

BIG-STREAM

A Framework for Digitisation in Africa's Circular Plastic Economy

Celine Ilo, Muyiwa Oyinlola and Oluwaseun Kolade

1 Introduction

According to United Nations (2021), the world's population is predicted to rise to 8.5 billion by 2050 and 11.2 billion by year 2100. This rapid increase in population coupled with the versatility of plastics to be adopted in various sectors of society (Mrowiec, 2018) has resulted in corresponding increases in the demand for natural resources such as salt, crude oil, natural gas, cellulose and coal required for the production of plastics (Plastics Europe, 2022). This translates to major strain on the earth's natural resources as a result of increased consumption of non-renewable fossil-based materials (Payne and Jones, 2021).

Poor waste management practices across the globe have resulted in severe consequences such as pollution of freshwater resources, clogging waterways and permeating sub-aquatic space (Awoyera and Adesina, 2020). Approximately 4.8–12.7 million tonnes of waste is expelled into water bodies from coastal areas every year (Conkle et al., 2017; Mrowiec, 2018). Furthermore, it has been reported that plastics and microplastics (plastics considered to be smaller than 5 mm) currently account for a reasonable proportion of marine debris. This is alarming as microplastics pose a threat to the sustenance of life underwater Given that smaller sea creatures and those in their formation stages can easily ingest these materials, thus introducing microplastics into oceanic food chains (Conkle et al., 2017). In addition, terrestrial biodiversity is threatened with the risk of extinction as a result of discharges emanating from toxic elements constituting plastic wastes, which saturate and pollute the ecosystem. Improper plastic waste management also poses far-reaching threats to public health as microplastics are taken up through air inhalation, ingestion or absorption when plastic wastes are being incinerated in some communities. The particles released during this activity

DOI: 10.4324/9781003278443-14

can be inhaled in the air, ingested when they settle on drinking water or absorbed through chemical transfer in food types consumed by humans (Marsden et al., 2019; Silva et al., 2022).

The non-biodegradable characteristic of plastics is attributable to its heavy molecular weight; hence, if not managed adequately at end of life, these will remain on the earth's surface for many years without decomposing or disintegrating (Sharuddin, Abnisa and Daud, 2016). This further underlines a requirement for the development of a sustainable solution aimed at the disruption of the prevalent linear economy for plastics and solid waste management in Africa.

Several scholars (e.g., Bakker et al., 2014; Rashid et al., 2013) posit that it is possible to address the global problem of poor plastic waste management by establishing a holistic system governed by the principles of the circular economy (CE), which regulates all phases in the plastic value chain (Kaur et al., 2018; Mrowiec, 2018). This system is referred to as the "Circular Plastic Economy (CPE)", and its rationale is hinged on transforming the methods of designing, producing and using plastic materials. In other words, the CPE aims to facilitate extended service life, value recovery and ecological compatibility for plastic resources. It entails a fundamental rethink of design and production approach which culminates in a closed-loop system for the life cycle of plastic resources (Payne and Jones, 2021).

Experts have highlighted the central roles of digital tools for the enhancement and efficacy of the CPE (Barrie et al., 2022; Oyinlola et al., 2022b; Rajput and Singh, 2019). Digital technologies enable strategic monitoring, predictive investigation, increased system performance and traceability through the material life cycle (Chauhan et al., 2019). Similarly, the efficient use of resources facilitated through data-informed regenerative designs improves the environmental and economic sustainability of plastic products. Instructional and predictive machine learning insights can hence be used to tailor the production processes as well as constituents of eco-friendly products (Bressanelli et al., 2018a; Garcia-Muiña et al., 2019).

Therefore, this chapter reviews the intersection between modern digital technologies and the CPE. It examines various models for optimising digital technologies for systemic changes in ecosystems. This leads to the conceptualisation of a framework for the digitisation of Africa's CPE. Accordingly, this chapter contributes to the body of literature as it targets the design of a holistic system for the intersection of digital innovations inspired by a significant range of digital functions and a CPE for Africa.

2 A Digitally Enabled CPE

Emerging digital technologies present great opportunities for the revolutionisation of critical sectors of the global economy. Digital technologies include Internet of things (IoT), artificial intelligence (AI), mobile applications, virtual reality (VR), augmented reality (AR), cloud computing, three-dimensional (3D) printing, geographic information systems (GIS) and remote sensing, blockchain technology and big data analytics (BDA). The integration of these digital technologies in the CE will enable the development of innovations addressing various social and economic issues currently experienced in different sectors and parts of the world (Oyinlola et al., 2022). In addition, digital innovations will allow for a seamless transition from the contemporary linear value network into a CE for plastic resources, as it fosters a shift from unsustainable methods of material sourcing, production and consumption (Liu et al., 2022). This is required to effectively address the plethora of ecological and climate-related problems plaguing our planet in recent times. Consequently, scholars have argued that accelerating the global shift to a CE is firmly tied to digitalisation (Ajwani-Ramchandani et al., 2021; Chauhan et al., 2022; Ingemarsdotter et al., 2019). Researchers have examined various technologies, for example, the application of AI as a digital tool capable of executing tasks in a manner synonymous to that of the human intellect in information assimilation and reasoning (Wilts et al., 2021). Digital innovations can be instrumental for the implementation of CE principles in various industries. As an illustration, the flow of products can be tracked by manufacturers' post-consumption in order to retrieve components and valuable parts for regeneration and design of value-added products (Lopes de Sousa Jabbour et al., 2018). Similarly, other scholars have shown that 3D printing can accelerate the transition to a CE (Oyinlola et al., 2023). Digital tools can be applied across the entire circular plastic value chain.

At present, the intersection of digital tools and the CE can be seen as a burgeoning field of research as there are a limited number of studies in this area. Recently developed literature draws upon ideas and analyses from domains such as competition-led sustainability in businesses, i.e., product service systems (PSS), industrial ecology and sustainable supply chain logistics (Pagoropoulos et al., 2017). Interestingly, conceptual research and reviews constitute the bulk of existing works as there are inadequate empirical studies illuminating the use of digital technology within the spheres of a CE, especially in developing regions like Africa. With the concept of a CE being often considered alongside other notions like decentralised manufacturing (Moreno and Charnley, 2016; Srai et al., 2016) and enterprise systems, some may argue that the area is still at a "pre-paradigmatic" stage (Pagoropoulos et al., 2017; Weichhart et al., 2016), such that it must be developed, while tailored to individual relevant disciplines.

According to recent estimates, 1 million plastic bottles are manufactured every minute, with single-use plastics accounting for 47% of total garbage (Fagnani et al., 2021; Payne et al., 2019). A sustainable plastic economy cannot be accomplished simply by renewable feedstock; there is a necessity for supplementation by extensive sustainable waste management strategies. This requires several digitally enhanced material recovery infrastructures in order to manage the massive amounts of plastic garbage produced per time and minimise any leakages from the sustainable network.

The fundamentals of a CE include eco-efficiency, material collection, sorting and recycling, sustainable design, production and redesign, life cycle assessment, cleaner production, carbon footprint reduction as well as other sustainable practices (Qi et al., 2016). Consequent to their multi-functional and long-lasting nature, the resourceful management of plastic products alongside the various processes involved in absolute value extraction from plastic wastes will be hardly achievable without digital technologies.

However, effective uptake of digital technologies for the CPE has been hampered by a number of barriers in Africa. These include inadequate information on how material resources and products traverse through the plastic value chain as well as their activities through their service life which will provide necessary details on their degradation processes and catalysts (Foschi et al., 2020). Another consideration is the lack of technological expertise for sustainable product design and how this expertise can be well inculcated into individual product development processes and projected service stages (Foschi et al., 2020). This will play a significant role in influencing general stakeholder (resource extraction companies, producers, manufacturers, retailers, customers and recyclers) behaviour in terms of levels of readiness and willingness to adopt required sustainable practices (Solomon and van Klyton, 2020). Dmitriev (2019) in a study enunciating the introduction of technologies for the logistics systems underlined challenges due to the lack of adequately defined legal framework, as well as the technical reticence of transport and logistics businesses to use modern digital technologies in the delivery of commodities. Therefore, it is impossible to disregard existing political and regulatory constraints such as the lack or misalignment of incentives, the absence of support from governmental institutions and hesitation on the path of business owners and product manufacturers (Bocken et al., 2016; Bressanelli et al., 2018b; Schirmeister and Mülhaupt, 2022; Schroeder et al., 2023). Foschi et al. (2020) further described how the public–private governance model, coupled with the growing number of disposal consortia and platforms, contributes challenges in product tracking. Olukanni et al. (2018) identified installation costs of a traditional material recovery facility(s) (MRFs) in low-income countries as well as a lack of significant technical skills, as a key impediment to the operationalisation of a CE for plastics.

A fundamental challenge for transitioning to a CPE in Africa is that the key actors (technical facilities, research bodies and governmental institutions) typically operate in silos (Oyinlola et al., 2022) with no strategic synergy and integration of approaches and methods by which the digital technologies are deployed to address various aspects of the CPE, such as collection, separation, sorting, sanitisation and recycling, to mention a few (Kolade et al., 2022; Olukanni et al., 2018; Oyinlola et al., 2022b). For example, many technology-driven initiatives and start-ups for the CPE in Africa have been seen to operate individually, effectively disconnected from one another (Oyinlola et al., 2022b). This makes it difficult to achieve significant changes as should be seen with a functional CPE. An efficient CPE across Africa will be unachievable with the current system of things which is characterised by the absence of unified participation of pertinent stakeholders (Awoyera and Adesina, 2020). As such, actualising the CPE calls for an amalgamation of consistent inputs from the diverse stakeholders involved. The CPE will benefit from a well-defined systemic change giving rise to a significant shift in societal values and norms (Chizaryfard et al., 2021).

Systemic changes that will accelerate the transition to a CPE cannot be facilitated by isolated digital innovations and disjointed CE strategies. There is a need for synergistic transformations across the entire value chain through the integration of multiple digital tools, strategically tailored to prevent leakage from the plastic value chain as well as track material flows (Truffer et al., 2008). In order to achieve this goal, a system thinking approach must be adopted. In a study highlighting the significance of digital technologies in the CE, Pagoropoulos, Pigosso and McAloone (2017) asserted that they have empowered the formulation of multiple PSS in the field of business. The concept of PSS is synonymous to the CE as it promotes a shift in business focus from selling things to selling utility via a combination of products and services that satisfies the same set of customer demands with less environmental effects (Lewandowski, 2016). Pagoropoulos, Pigosso and McAloone (2017) further evaluated the efficacy of digital technologies in the CE using a three-layer architectural framework namely, data collection, data integration and data analysis. Seven digital tools were identified and grouped into each layer based on individual functions: for data collection, radio frequency identification (RFID) and IoT; for data integration, relational database management systems (RDBMS), product life cycle management (PLM) systems and AI; and for data analysis, machine learning and BDA. An evaluation of the framework depicts that digital technologies play an essential role towards the CE by acting as a critical enabler in the optimisation of forward material flows and expedition of reverse material flows (Pagoropoulos, Pigosso and McAloone, 2017).

Chauhan, Sharma and Singh (2019) employed the situation, actor, process–learning, action, performance (SAP–LAP) interconnection model to examine the applicability of Industry 4.0 mechanisms in resolving difficulties in existing CE business models. This was achieved by analysing the cross-interaction and self-interaction linkages between the various components of the SAP–LAP framework, thus integrating both CE and Industry 4.0 streams in order to ascertain how issues regarding the CE parameters can be tackled. Research findings based on developed toolkits suggest that as regards the CE, senior managers (actors) have the most influence on the integration of Industry 4.0 to achieve sustainability. Additionally, smart technologies like IoT and cyber physical systems account for the most important Industry 4.0 activities that encourage the enhancement of CE performance metrics. However, the shortcoming of this research work is that the identification of ties between the main components of the SAP–LAP framework is based on the personal judgement of various experts; thus, it is susceptible to the writer's individual bias. The study will benefit from empirical validation and real-world implementation.

Furthermore, Liu et al. (2022) conducted a systematic review to inform the development of a framework – digital technologies for the circular economy(DT4CE). This framework is used to ascertain which digital functions are most relevant in the realisation of a functional CE. The study identified 13 different digital functions (Auto-plan, Auto-control, Sort and Classify, Optimise, Innovate, Forecast, Connect, Assess, Detect, Track and Trace, Monitor, Share and Collect) to be most effective in driving material circularity in line with the CE principles. It further hinges on seven mechanisms (Recycle, Repurpose, Remanufacture, Repair, Reuse, Reduce and Rethink and Refuse) for the implementation of the selected functions towards the enhancement of CE strategies. The framework also examines specifications and combinations of digital functions and CE strategies that have been widely studied, thereby revealing levels of technology maturity and existing gaps for application in the CE. Albeit the study is limited by its emphasis on just three digital technologies, namely IoT, BDA and AI. Reviews on a larger variety of CE technologies would offer greater understanding of the pertinence of digital technologies in the circular economy (CE-DT) integration.

Cwiklicki and Wojnarowska (2020) compared technologies such as AI, robotics, the IoT, autonomous vehicles, 3D printing, nanotechnology and biotechnology using the ReSOLVE model, 3R strategy and three other concepts. They concluded that the IoT and BDA were the most promising Industry 4.0 digitalisation tools for the CE. Ingemarsdotter et al. (2019), in their model, incorporated the 3R strategy with three operational strategies to point out the potentials of IoT. Tracking, monitoring, control, optimisation and design evolution were identified as main IoT capabilities, while circular in-service strategies are efficiency in use, increased utilisation and product service life extension. Circular looping strategies include reuse, remanufacturing and recycling. Furthermore, case studies on digital tools such as IoT, big data and data analytics were categorised by Kristoffersen et al. (2020) using the circular strategies scanner by Blomsma et al. (2019) which entails a comprehensive multilayered strategy mapping in accordance with the 9R strategies formulated by Potting et al. (2017).

3 A Framework for a Digitally Enabled CPE

As highlighted in the previous section, the application of multiple digital technologies in tandem, can perform a variety of essential functions. However, most of the studies have been focused on the Global West, with only a few fragmented studies focused on Africa which are not comprehensive enough to provide understanding on how digital technologies can accelerate a systemic shift in Africa's current plastic value chain (Aristi Capetillo, 2021). Desmond and Asamba (2019) also noted that African case studies stay "hidden" as they are yet to be documented through academic research. Therefore, this chapter makes a contribution by drawing on a review of the extant literature to develop

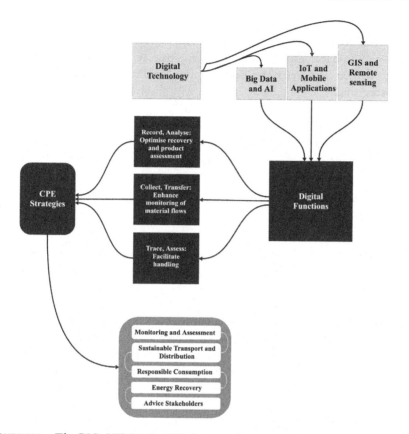

FIGURE 11.1 The BIG-STREAM CPE framework

a framework, the BIG-STREAM framework, which can accelerate the CPE transition in Africa.

This framework focuses on the following three main areas:

1. The core functions of digital technologies that are relevant to the CPE transition in Africa; Digital functions refer to using specific technologies to perform smart actions (Allmendinger and Lombreglia, 2005).
2. The strategies to be prioritised for the CPE transition in Africa.
3. The mechanisms by which highlighted digital functions can be leveraged for effective implementation of the CPE.

The elements of the BIG-STREAM framework (shown in Figure 11.1) are discussed below:

- B – big data and AI
- I – IoT and mobile applications

- G – GIS and remote sensing
- ST – sustainable transport and distribution
- R – responsible consumption; use, reuse, repair, remanufacture and repurpose, and recycle and recovery; identification, collection, separation, sorting, reprocessing
- E – energy recovery
- A – advice stakeholders; refrain, reuse, separate and garner
- M – monitoring and assessing of waste management systems for improved and more sustainable future product designs.

3.1 Digital Functions

3.1.1 BDA and AI

BDA is referred to as the analysis of large data sets using a variety of cutting-edge methodologies in order to draw inferences and valuable conclusions (Ghasemaghaei et al., 2015; Oztemel and Gursev, 2018). Similarly, in the scope of a CPE for diverse settings like Africa, BDA embodies a viable tool for leveraging information from multiple systems of record, such as sensors and IoT, to enhance decision-making. It is pertinent to highlight that big data is not generally treated as a concept in and of itself but rather as a method for analysing large amounts of data gathered from various data sources (Abideen et al., 2021). This is particularly relevant taking into consideration the regional and ecological complexity of Africa (Olukanni et al., 2018). AI has also lately received more attention in studies pertaining to the CE. It enables rapid and adaptive learning for data analysis (Haenlein and Kaplan, 2019; Kristoffersen et al., 2020), allowing for faster and more dynamic operations based on larger data sets and therefore opening up new opportunities for CE implementations (Ellen Macarthur Foundation, 2013). Relative to plastic waste management, the amalgamation of BDA and AI will enhance the tracing and sharing of material flows from the design and production to the end-of-service life stages which invariably facilitates waste recovery and connectivity between waste reduction practices. The integration of AI further introduces a smart and agile learning interface for data analysis and allows faster and more adaptable actions using large and dispersed data sets (Esmaeilian et al., 2018; Kristoffersen et al., 2020). IoT-affiliated technologies such as VR and AR are instrumental in the area of educating CPE stakeholders such as potential customers and governmental bodies on the imminence of environmental disaster if their quota is not rendered towards the CPE as well as the myriad of benefits to be recovered from existing plastic waste materials.

3.1.2 IoT and Mobile Applications

Digital tools such as IoT which are known to function by virtue of linking objects to enable the collection and transfer of information (Ghasemaghaei et al., 2015). On the other hand, mobile and web applications for cell phones and other devices have become vital tools for a wide range of applications in the context of the CE (Faria et al., 2020). They rely on various methods of connectivity such as near-field communication (NFC), Bluetooth low energy (BLE) and Wi-Fi, for short-range communications and technologies and general packet radio service (GPRS), universal mobile telecommunications system (UMTS) and 3G/4G/5G to support long-range connections (Marques et al., 2019). Thereby enabling practices such as electronic commerce and product exchange, based on the geolocation of neighbouring users, and material upcycling (where it serves as a traditional social network for acquiring used items), which are required in the handling of plastic resources and waste materials (Agrebi and Jallais, 2015).

In terms of the identified life cycle phases in a CE for plastics, sensor technologies such as RFIDs, which is a major tool on which the IoT technology depends, in conjunction with mobile applications as principal interface, may assist in closing the material loop for plastics. Synchronous incorporation of both digital functions will promote system resilience for the CPE within Africa as it presents a medium for seamless communication between CPE actors. In this scenario, the key emphasis is on the material "End of Life" (EoL) and its relationship with sustainable manufacturing and product redesign. One famous example is the employment of RFIDs in the development of useful information on how the customer or client handled the product; hence, the incorporation of functions such as data collection will facilitate monitoring of plastic material flows through their life cycle (Faria et al., 2020).

3.1.3 GIS and Remote Sensing

The development of new digital technologies like remote sensing and GIS have made municipal waste management assessments seamless to conduct in recent years. Critical processes involved in the management of solid wastes like plastics, including capturing, waste handling and the transmission of necessary information in a timely and error-free manner, have been enhanced through the employment of these technologies. These tools may also be used to gather information directly from a distant site at a reasonable cost. Remote sensing technologies are also applied in landfill and trash bin placement, as well as assessing the environmental effect of buried garbage. Techniques have also been used for locating landfills and waste stockpiles for disposal, as well as evaluating the ecological effects of buried debris (Singh, 2019). This technology will be instrumental in the operationalisation of the recycle and recovery strategy (identification, collection, separation, sorting and material reprocessing) for plastic material circularity, as its monitoring capabilities will facilitate ease of identification and collection of plastic waste for effective recycling and upcycling.

3.2 Strategies

3.2.1 Sustainable Transport and Distribution

The development of eco-packaging designs and environmentally friendly plastic substitution to facilitate material flows for plastic containers, and enhance packaging inventory is crucial as plastic packaging accounts for a huge percentage of overall plastic waste products in the environment today. This is also owing to the fact that in the contemporary mechanisms, they are designed to be single use, thereby causing their distribution to imply environmental depletion (Esmaeilian et al., 2018). Studies performed by Galindo et al. (2021) portrayed how the processing of raw materials to completed goods, raw material deployment and final product distribution can be sustainably optimised regardless of the extensivity and possible disjunction of existing suppliers' network, product units and consumers. The model targeted the reduction of overall costs, which include raw material acquisition costs, manufacturing expenses and transportation costs, leveraging information such as raw material supply, raw material prices, raw material needs, production capacities, production expenses, raw material and final product conversions, customer demand and transportation costs provided by the plastic manufacturing firm.

3.2.2 Responsible Production and Consumption

They predominantly include, but are not limited to, the Reduce, Reuse and Recycle strategies commonly referred to as the 3Rs of the CPE (Blomsma et al., 2019). Extrapolation of the 3Rs, in attempts to assess the CE strategies through a more comprehensive and circular perspective (cradle to cradle), led to the 9R strategy, R_0 – Refuse, R_1 – Rethink, R_2 – Reduce, R_3 – Reuse, R_4 – Repair, R_5 – Refurbish, R_6 – Remanufacture, R_7 – Repurpose, R_8 – Recycle and R_9 – Recover, which was developed by Potting et al. (2017).

Innovative solutions such as the reuse, repair, remanufacturing and repurposing of plastic products should be employed to prevent plastics and microplastics from leaking into the environment, reaching and settling into water bodies. The effective execution of these practices is also facilitated by the use of alternative sustainable feedstock for plastic production, compared to non-renewable options (Mrowiec, 2018).

Adopting clean methods in the sourcing of plastic feedstock is essential for maintaining an environmentally sustainable plastic economy. Bioplastics such as polylactic acid was introduced and promoted by Djukić-Vuković et al. (2019) as a potential substitute to improve the environmental efficiency of plastics, whereas Walker and Rothman (2020) claimed that the precise environmental impacts of bioplastics are yet to be defined. Payne et al. (2019) explored polylactic acid waste management strategies. Plant-based plastic necessitates the use of fertilisers, pesticides and land, accounting for further consumption of natural capital and the disruption of soil fauna (Atwood et al., 2018). Bioplastics can be made from waste, such as food waste, to produce polyhydroxyalkanoates (Rai et al., 2019), which has a lower environmental footprint. However, the economic viability of pre-treatment and the spillover impacts (requirement for food waste or unstable supply) must be carefully evaluated. Another option is biodegradable plastics (petroleum based with additives). However, they can still produce debris and pollution in a case where they are not properly collected, as they are not fully decayed under all conditions. Ultimately, the issues raised by plastic consumption would perpetuate if the waste management supply chain is not well developed and strongly adhered to (Klemeš et al., 2020).

The revision of current design and manufacturing techniques will allow for higher plastic recycling rates in all major applications. A myriad of plastic products, in fact, cannot be reused or recycled, in some cases relative to their method of initial design as well as material type. Thus, product redesign involves the utilisation of alternative available materials, for instance, the employment of natural alternatives to plastic microbeads in beauty products. In the same vein, it is important to consider the design of plastic products without the addition of toxic chemicals and colorants as this could result in ecological and health problems as well as minimises product capacity in secondary applications (Brink et al., 2018). This strategy will benefit from compliance on the path of managers in the adoption of eco-friendly product designs in a bid to avert the risk associated with customer's unwillingness to abandon traditional products (De Jesus and Mendonça, 2018; Ritzén and Sandström, 2017).

Increased rates of plastic recycling cut down reliance on the importation of fossil fuels and reduce CO_2 emissions. The processes of gathering, separation, sorting and recycling of plastic waste materials contribute to job opportunities and flexible income generation (Klemeš et al., 2020). It is a vital phase of the circular plastics loop and a determinant for the materialisation of plastic circularity.

3.2.3 Energy Recovery

Recycling and energy recovery are the final lines of defence in reshaping the linear system into a CE because they allow plastic products to have a longer lifespan and maintain resources in use for as long as possible, enhancing the sustainable management of post-consumer plastics which are at their end of life (Klemeš et al., 2020). The ultimate place of non-recyclable plastic should be incineration. Liu et al. (2022) argued that several factors influence the choice of an energy-based disposal technique, including energy efficiency, technical specifications, environmental laws, social acceptance and responsibility. Incineration, autoclaving, microwaving, plasma treatment, chemical treatment and steam treatment are examples of traditional techniques.

3.2.4   Advice Stakeholders: Refrain, Reuse, Separate and Garner

According to Wichai-utcha and Chavalparit (2018), fundamental hindrances to the recycling of plastic waste materials are three major factors. They include human behaviour, i.e., lack of awareness on plastic recycling processes and the various types of plastics available; regulations (presence of incentives, plastic resin identification codes and eco-labels on products, etc.); and, lastly, recovery infrastructures (availability and access to collection bodies, efficiency of operations and running costs). These factors further highlight the cost intensiveness of effective plastic waste management and the need for integration of digital tools to close existing gaps by increasing the awareness of plastic users, knowledge proliferation, encouraging waste reductions, government intervention and promoting flow and connectivity in the plastics economy (Olukanni et al., 2018). This further allows for the advancement of scientific knowledge and extensive adoption of sustainable practices (Bucknall, 2020; Liu et al., 2022).

3.2.5   Monitoring and Assessing Waste Management Systems

The incorporation of digital functions is pertinent in this area, various initiatives have recorded the use of tools like blockchain technology. The monitoring and evaluation of waste management systems is invaluable to the proliferation of the CPE concept. In order to collect longitudinal data on waste operations, blockchains are characterised as a data ledger, and as such, one of its basic features is the logging of events and transactions in blocks, allowing the provenance of resources and wastes to be made available and public, if required. This data is utilised in the monitoring and improvement of the efficacy and efficiency of waste management procedures (Steenmans et al., 2021).

In this study, the decision was made to harness three major technologies for optimising the CPE and some in clusters as they share an intersection of functions relevant to the proposed CE strategies. They include big data and AI, IoT and mobile applications, and GIS and remote sensing, thereby constituting the "BIG-STREAM" framework.

4   Conclusion

Given the environmental and public health challenges posed by plastics, the transition to a CPE in Africa is now imperative. This chapter illustrates that despite the potential for digital technologies to accelerate the transition, there is need for strategic synergy and integration of approaches and methods. Therefore, this chapter adopts a system thinking approach to develop the BIG-STREAM framework, which brings together digital functions, strategies and mechanisms for digital technologies to address various aspects of the CPE. This chapter also stresses the necessity for the incorporation of digital innovations to foster a sustainable and resource-efficient plastic value chain which will function in lockstep with a proactive atmosphere of collaboration among stakeholders. The framework underlines practical as well as research implications which includes a requirement for the stringent reform of business practices and a change of social behaviour alongside further research on how digitally optimised frameworks such as this could be closely tailored to specific plastic-affiliated industries, organisations and businesses. This will allow for clearer insights on how their management processes can be reconfigured as well as expose new opportunities for perpetual development. Finally, the limitation of this study includes that the digital functions discussed are nascent, and the CPE strategies were established based on a review of conceptual studies or systematic reviews of current relevant academic resources; hence, there is requirement for further validation of ideas through quantitative and practical case studies for a CPE.

Acknowledgement

This work was supported by the Sustainable Manufacturing and Environmental Pollution (SMEP) programme funded with UK aid from the UK government Foreign, Commonwealth and Development Office (FCDO) (IATI reference number GB-GOV-1-30012).

References

Abideen, A.Z., Pyeman, J., Sundram, V.P.K., Tseng, M.L., Sorooshian, S., 2021. Leveraging capabilities of technology into a circular supply chain to build circular business models: A state-of-the-art systematic review. *Sustainability* 13, 8997. https://doi.org/10.3390/SU13168997

Agrebi, S., Jallais, J., 2015. Explain the intention to use smartphones for mobile shopping. *J. Retail. Consum. Serv.* 22, 16–23. https://doi.org/10.1016/J.JRETCONSER.2014.09.003

Ajwani-Ramchandani, R., Figueira, S., Torres de Oliveira, R., Jha, S., Ramchandani, A., Schuricht, L., 2021. Towards a circular economy for packaging waste by using new technologies: The case of large multinationals in emerging economies. *J. Clean. Prod.* 281, 125139. https://doi.org/10.1016/J.JCLEPRO.2020.125139

Allmendinger, G., Lombreglia, R., 2005. Four strategies for the age of smart services. *Harvard Bus. Rev.*, 76, 131–145.

Aristi Capetillo, A., 2021. Circular economy in a plastic world: How can emerging technologies enable the transition? Master's Thesis.

Atwood, L.W., Mortensen, D.A., Koide, R.T., Smith, R.G., 2018. Evidence for multi-trophic effects of pesticide seed treatments on non-targeted soil fauna. *Soil Biol. Biochem.* 125, 144–155. https://doi.org/10.1016/J.SOILBIO.2018.07.007

Awoyera, P.O., Adesina, A., 2020. Plastic wastes to construction products: Status, limitations and future perspective. *Case Stud. Constr. Mater.* 12, e00330. https://doi.org/10.1016/j.cscm.2020.e00330

Bakker, C., Wang, F., Huisman, J., Den Hollander, M., 2014. Products that go round: Exploring product life extension through design. *J. Clean. Prod.* 69, 10–16. https://doi.org/10.1016/J.JCLEPRO.2014.01.028

Barrie, J., Anantharaman, M., Oyinlola, M., Schröder, P., 2022. The circularity divide: What is it? And how do we avoid it? *Resour. Conserv. Recycl.* 180, 106208. https://doi.org/10.1016/J.RESCONREC.2022.106208

Blomsma, F., Pieroni, M., Kravchenko, M., Pigosso, D.C.A., Hildenbrand, J., Kristinsdottir, A.R., Kristoffersen, E., Shabazi, S., Nielsen, K.D., Jönbrink, A.K., Li, J., Wiik, C., McAloone, T.C., 2019. Developing a circular strategies framework for manufacturing companies to support circular economy-oriented innovation. *J. Clean. Prod.* 241, 118271. https://doi.org/10.1016/J.JCLEPRO.2019.118271

Bocken, N.M.P., De Pauw, I., Bakker, C., Van Der Grinten, B., 2016. Product design and business model strategies for a circular economy. *J. Ind. Prod. Eng.* https://doi.org/10.1080/21681015.2016.1172124

Bressanelli, G., Adrodegari, F., Perona, M., Saccani, N., 2018a. Exploring how usage-focused business models enable circular economy through digital technologies. *Sustainability* 10, 639. https://doi.org/10.3390/SU10030639

Bressanelli, G., Adrodegari, F., Perona, M., Saccani, N., 2018b. The role of digital technologies to overcome circular economy challenges in PSS business models: An exploratory case study. *Procedia Cirp.* 73, 216–221.

Brink, P. ten *et al.* 2018. Circular economy measures to keep plastics and their value in the economy, avoid waste and reduce marine litter. Available at: www.econstor.eu/han dle/10419/173128

Bucknall, D.G., 2020. Plastics as a materials system in a circular economy. *Philos. Trans. R. Soc. A* 378, 20190268.

Chauhan, C., Parida, V., Dhir, A., 2022. Linking circular economy and digitalisation technologies: A systematic literature review of past achievements and future promises. *Technol. Forecast. Soc. Change* 177, 121508. https://doi.org/10.1016/J.TECHF ORE.2022.121508

Chauhan, C., Sharma, A., Singh, A., 2019. A SAP–LAP linkages framework for integrating Industry 4.0 and circular economy. *Benchmarking* 28, 1638–1664. https://doi.org/10.1108/BIJ-10-2018-0310/FULL/PDF

Chizaryfard, A., Trucco, P., Nuur, C., 2021. The transformation to a circular economy: framing an evolutionary view. *J. Evol. Econ.* 31, 475–504. https://doi.org/10.1007/s00191-020-00709-0

Conkle, J.L., Báez Del Valle, C.D., Turner, J.W., 2017. Are we underestimating microplastic contamination in aquatic environments? *Environ. Manag.* 61, 1–8. https://doi.org/10.1007/S00267-017-0947-8

Cwiklicki, M., Wojnarowska, M., 2020. Circular economy and Industry 4.0: One-way or two-way relationships? *Eng. Econ.* 31, 387–397. https://doi.org/10.5755/J01. EE.31.4.24565

De Jesus, A., Mendonça, S., 2018. Lost in transition? Drivers and barriers in the eco-innovation road to the circular economy. *Ecol. Econ.* 145, 75–89.

Desmond, P., Asamba, M., 2019. Accelerating the transition to a circular economy in Africa: Case studies from Kenya and South Africa. *Circ. Econ. Glob. South* 152–172. https://doi.org/10.4324/9780429434006-9

Djukić-Vuković, A., Mladenović, D., Ivanović, J., Pejin, J., Mojović, L., 2019. Towards sustainability of lactic acid and poly-lactic acid polymers production. *Renew. Sustain. Energy Rev.* 108, 238–252. https://doi.org/10.1016/J.RSER.2019.03.050

Dmitriev, A.V., 2019. Digital technologies of transportation and logistics systems visibility. *Strateg. Decis. risk Manag.* 10, 20–26. https://doi.org/10.17747/2618-947X-2019-1-20-26

Ellen Macarthur Foundation, 2013. Towards the Circular Economy: Economic and Business Rationale for an Accelerated Transition. Cowes.

Esmaeilian, B., Wang, B., Lewis, K., Duarte, F., Ratti, C., Behdad, S., 2018. The future of waste management in smart and sustainable cities: A review and concept paper. *Waste Manag.* 81, 177–195. https://doi.org/10.1016/J.WASMAN.2018.09.047

Fagnani, D.E., Tami, J.L., Copley, G., Clemons, M.N., Getzler, Y.D.Y.L., McNeil, A.J., 2021. 100th anniversary of macromolecular science viewpoint: redefining sustainable polymers. *ACS Macro Lett.* 10, 41–53. https://doi.org/10.1021/ACSMACROL ETT.0C00789/SUPPL_FILE/MZ0C00789_LIVESLIDES.MP4

Faria, R., Lopes, I., Pires, I.M., Marques, G., Fernandes, S., Garcia, N.M., Lucas, J., Jevremovic, A., Zdravevski, E., Trajkovik, V., 2020. Circular economy for clothes using web and mobile technologies—A systematic review and a taxonomy proposal. *Information* 11, 161. https://doi.org/10.3390/INFO11030161

Foschi, E., D'Addato, F., Bonoli, A., 2020. Plastic waste management: a comprehensive analysis of the current status to set up an after-use plastic strategy in Emilia-Romagna Region (Italy). *Environ. Sci. Pollut. Res.* 28, 24328–24341. https://doi.org/10.1007/S11 356-020-08155-Y

Galindo, A.M.O., Dadios, E.P., Billones, R.K.C., Valenzuela, I.C., 2021. Cost Optimization for the Allocation, Production, and Distribution of a Plastic Manufacturing Company Using Integer Linear Programming. 2021 IEEE 13th Int. Conf. Humanoid, Nanotechnology, Inf. Technol. Commun. Control. Environ. Manag. HNICEM 2021. https://doi.org/10.1109/HNICEM54116.2021.9731804

Garcia-Muiña, F.E., González-Sánchez, R., Ferrari, A.M., Volpi, L., Pini, M., Siligardi, C., Settembre-Blundo, D., 2019. Identifying the equilibrium point between sustainability goals and circular economy practices in an Industry 4.0 manufacturing context using eco-design. *Soc. Sci.* 8, 241. https://doi.org/10.3390/SOCSCI8080241

Ghasemaghaei, M., Hassanein, K., Turel, O., 2015. Impact of Data Analytics on Organizational Performance Impacts of Big Data Analytics on Organizations: A Resource Fit Perspective Emergent Research Forum Papers 1–2.

Haenlein, M., Kaplan, A., 2019. A brief history of artificial intelligence: On the past, present, and future of artificial intelligence. *California Manag. Rev.* 61, 5–14. https://doi. org/10.1177/0008125619864925

Ingemarsdotter, E., Jamsin, E., Kortuem, G., Balkenende, R., 2019. Circular strategies enabled by the Internet of Things—A framework and analysis of current practice. *Sustainability* 11, 5689. https://doi.org/10.3390/SU11205689

Kaur, G., Uisan, K., Ong, K.L., Ki Lin, C.S., 2018. Recent trends in green and sustainable chemistry & waste valorisation: Rethinking plastics in a circular economy. *Curr. Opin. Green Sustain. Chem.* 9, 30–39. https://doi.org/10.1016/j.cogsc.2017.11.003

Klemeš, J.J., Van Fan, Y., Jiang, P., 2020. Plastics: Friends or foes? The circularity and plastic waste footprint. *Energy Sources Part A* 43, 1549–1565. https://doi.org/10.1080/ 15567036.2020.1801906

Kolade, O., Odumuyiwa, V., Abolfathi, S., Schröder, P., Wakunuma, K., Akanmu, I., Whitehead, T., Tijani, B., Oyinlola, M., 2022. Technology acceptance and readiness of stakeholders for transitioning to a circular plastic economy in Africa. *Technol. Forecast. Soc. Change* 183, 121954. https://doi.org/10.1016/J.TECHFORE.2022.121954

Kristoffersen, E., Blomsma, F., Mikalef, P., Li, J., 2020. The smart circular economy: A digital-enabled circular strategies framework for manufacturing companies. *J. Bus. Res.* 120, 241–261. https://doi.org/10.1016/J.JBUSRES.2020.07.044

Lewandowski, M., 2016. Designing the business models for circular economy—Towards the conceptual framework. *Sustainability* 8, 43.

Liu, Q., Trevisan, A.H., Yang, M., Mascarenhas, J., 2022. A framework of digital technologies for the circular economy: Digital functions and mechanisms. *Bus. Strateg. Environ.* 31, 2171–2192. https://doi.org/10.1002/bse.3015

Lopes de Sousa Jabbour, A.B., Jabbour, C.J.C., Godinho Filho, M., Roubaud, D., 2018. Industry 4.0 and the circular economy: A proposed research agenda and original roadmap for sustainable operations. *Ann. Oper. Res.* 270, 273–286. https://doi.org/10.1007/S10479-018-2772-8

Marques, G., Pitarma, R., Garcia, N.M., Pombo, N., 2019. Internet of Things architectures, technologies, applications, challenges, and future directions for enhanced living environments and healthcare systems: A review. *Electronic* 8, 1081. https://doi.org/10.3390/ELECTRONICS8101081

Marsden, P. *et al.* (2019). *Microplastics in drinking water.* Geneva: World Health Organization. Available at: https://library.wur.nl/WebQuery/wurpubs/553048 (Accessed: 18 March 2023).

Moreno, M., Charnley, F., 2016. Can re-distributed manufacturing and digital intelligence enable a regenerative economy? An integrative literature review. *Smart Innov. Syst. Technol.* 52, 563–575. https://doi.org/10.1007/978-3-319-32098-4_48/COVER

Mrowiec, B., 2018. Plastics in the circular economy (CE). *Ochr. Sr. i Zasobow Nat.* 29, 16–19. https://doi.org/10.2478/OSZN-2018-0017

Olukanni, D.O., Aipoh, A.O., Kalabo, I.H., 2018. Recycling and reuse technology: Waste to wealth initiative in a private tertiary institution, Nigeria. *Recycling* 3, 44. https://doi.org/10.3390/RECYCLING3030044

Oyinlola, M., Kolade, O., Schroder, P., Odumuyiwa, V., Rawn, B., Wakunuma, K., Sharifi, S., Lendelvo, S., Akanmu, I., Mtonga, R., Tijani, B., Whitehead, T., Brighty, G., Abolfathi, S., 2022. A socio-technical perspective on transitioning to a circular plastic economy in Africa. *SSRN Electron. J.* https://doi.org/10.2139/ssrn.4332904.

Oyinlola, M., Okoya, S.A., Whitehead, T., Evans, M., Lowe, A.S., 2023. The potential of converting plastic waste to 3D printed products in Sub-Saharan Africa. *Resour. Conserv. Recycl. Adv.* 17, 200129. https://doi.org/10.1016/j.rcradv.2023.200129

Oyinlola, M., Schröder, P., Whitehead, T., Kolade, S., Wakunuma, K., Sharifi, S., Rawn, B., Odumuyiwa, V., Lendelvo, S., Brighty, G., Tijani, B., Jaiyeola, T., Lindunda, L., Mtonga, R., Abolfathi, S., 2022b. Digital innovations for transitioning to circular plastic value chains in Africa. *Africa J. Manag.* 8, 83–108. https://doi.org/10.1080/23322373.2021.1999750

Oztemel, E., Gursev, S., 2018. Literature review of Industry 4.0 and related technologies. *J. Intell. Manuf.* 31, 127–182. https://doi.org/10.1007/S10845-018-1433-8

Pagoropoulos, A., Pigosso, D.C.A., McAloone, T.C., 2017. The emergent role of digital technologies in the circular economy: A review. *Procedia CIRP* 64, 19–24.

Payne, J., Jones, M.D., 2021. The chemical recycling of polyesters for a circular plastics economy: Challenges and emerging opportunities. *ChemSusChem* 14, 4041–4070. https://doi.org/10.1002/cssc.202100400

Payne, J., McKeown, P., Jones, M.D., 2019. A circular economy approach to plastic waste. *Polym. Degrad. Stab.* 165, 170–181. https://doi.org/10.1016/J.POLYMDEGRADSTAB.2019.05.014

Plastics Europe (2022). *Enabling a sustainable future, Plastics Europe.* Available at: https://plasticseurope.org/ (Accessed: 18 March 2023).

Potting, J., Hekkert, M.P., Worrell, E., Hanemaaijer, A., 2017. Circular Economy: Measuring Innovation in the Product Chain. *Planbur. voor Leefomgeving.*

Qi, J., Zhao, J., Li, W., Peng, X., Wu, B., Wang, H., 2016. *Development of Circular Economy in China.* Research Series on the Chinese Dream and China's Development Path. Springer. https://doi.org/10.1007/978-981-10-2466-5

Rai, P.K., Lee, S.S., Zhang, M., Tsang, Y.F., Kim, K.H., 2019. Heavy metals in food crops: Health risks, fate, mechanisms, and management. *Environ. Int.* 125, 365–385. https://doi.org/10.1016/J.ENVINT.2019.01.067

Rajput, S., Singh, S.P., 2019. Connecting circular economy and industry 4.0. *Int. J. Inf. Manage.* 49, 98–113. https://doi.org/10.1016/J.IJINFOMGT.2019.03.002

Rashid, A., Asif, F.M.A., Krajnik, P., Nicolescu, C.M., 2013. Resource conservative manufacturing: An essential change in business and technology paradigm for sustainable manufacturing. *J. Clean. Prod.* 57, 166–177. https://doi.org/10.1016/J.JCLE PRO.2013.06.012

Ritzén, S., Sandström, G.Ö., 2017. Barriers to the circular economy – Integration of perspectives and domains. *Procedia CIRP* 64, 7–12. https://doi.org/10.1016/J.PRO CIR.2017.03.005

Sharuddin, S.D.A., Abnisa, F., Wan Daud, W.M.A., Aroua, M.K., 2016. A review on pyrolysis of plastic wastes. *Energy Convers. Manag.* 115, 38. https://doi.org/10.1016/j.enconman.2016.02.037

Schirmeister, C.G., Mülhaupt, R., 2022. Closing the carbon loop in the circular plastics economy. *Macromol. Rapid Commun.* 2200247. https://doi.org/10.1002/marc.202200247

Schroeder, P., Oyinlola, M., Barrie, J., Bonmwa, F., Abolfathi, S., 2023. Making policy work for Africa's circular plastics economy. *Resour. Conserv. Recycl.* 190, 106868. https://doi.org/10.1016/j.resconrec.2023.106868

Silva, D., Rocha-Santos, T.A.P., Adeniran, A.A., Shakantu, W., 2022. The health and environmental impact of plastic waste disposal in South African townships: A review. *Int. J. Environ. Res. Public Heal.* 19, 779. https://doi.org/10.3390/IJERPH19020779

Singh, A., 2019. Remote sensing and GIS applications for municipal waste management. *J. Environ. Manage.* 243, 22–29. https://doi.org/10.1016/j.jenvman.2019.05.017

Solomon, E.M., van Klyton, A., 2020. The impact of digital technology usage on economic growth in Africa. *Util. Policy* 67, 101104. https://doi.org/10.1016/J.JUP.2020.101104

Srai, J.S., Kumar, M., Graham, G., Phillips, W., Tooze, J., Ford, S., Beecher, P., Raj, B., Gregory, M., Tiwari, M.K., Ravi, B., Neely, A., Shankar, R., Charnley, F., Tiwari, A., 2016. Distributed manufacturing: Scope, challenges and opportunities. *Int. J. Prod. Res.* 54, 6917–6935. https://doi.org/10.1080/00207543.2016.1192302

Steenmans, K., Taylor, P., Steenmans, I., 2021. Blockchain technology for governance of plastic waste management: Where are we? *Soc. Sci.* 10, 434. https://doi.org/10.3390/SOCSCI10110434

Truffer, B., Voß, J.P., Konrad, K., 2008. Mapping expectations for system transformations: Lessons from sustainability foresight in German utility sectors. *Technol. Forecast. Soc. Change* 75, 1360–1372. https://doi.org/10.1016/J.TECHF ORE.2008.04.001

United Nations. 2021. *Peace, dignity and equality on a healthy planet, United Nations.* Available at: https://www.un.org/en/ (Accessed: 18 March 2023).

Walker, S., Rothman, R., 2020. Life cycle assessment of bio-based and fossil-based plastic: A review. *J. Clean. Prod.* 261, 121158. https://doi.org/10.1016/J.JCLE PRO.2020.121158

Weichhart, G., Molina, A., Chen, D., Whitman, L.E., Vernadat, F., 2016. Challenges and current developments for sensing, smart and sustainable enterprise systems. *Comput. Ind.* 79, 34–46. https://doi.org/10.1016/J.COMPIND.2015.07.002

Wichai-utcha, N., Chavalparit, O., 2018. 3Rs policy and plastic waste management in Thailand. *J. Mater. Cycles Waste Manag.* 21, 10–22. https://doi.org/10.1007/S10 163-018-0781-Y

Wilts, H., Garcia, B.R., Garlito, R.G., Gómez, L.S., Prieto, E.G., 2021. Artificial intelligence in the sorting of municipal waste as an enabler of the circular economy. *Resources* 10, 28. https://doi.org/10.3390/RESOURCES10040028

12

A PLASTIC DATA EXCHANGE PLATFORM FOR AFRICA'S CIRCULAR PLASTIC ECONOMY TRANSITION

Olawunmi Ogunde, Muyiwa Oyinlola and Stuart R. Coles

1 Introduction

Plastics have become an essential part of our day-to-day lives as they are used in product packaging, appliances, furniture, clothes, automotive applications and disposable single-use items (Soós *et al.*, 2021). Companies involved across the plastic supply chain are keen to highlight the benefit of these materials at any given time while end users are also keen to comment on how it prevents food waste (Paiho *et al.*, 2020) and its packaging advantages (Lopez-Aguilar *et al.*, 2022). Consequently, the versatility of plastics has resulted in the use of plastics across a wide range of industry applications (Thomas *et al.*, 2022). However, the lack of appropriate disposal (Wagh *et al.*, 2022) has now caused a wave of plastic pollution and has become a serious global concern (Soós *et al.*, 2021) worsened by growing production (Thomas *et al.*, 2022) and consumption (Ryberg *et al.*, 2019). For example, Dasgupta *et al.* (2022) reported that the use of plastics has grown at a rate of more than 8% per year since the end of the Second World War. Similarly, Geyer *et al.* (2017) estimated that 8,300 million tonnes of virgin material has been produced and approximately 6,300 million tonnes of plastic waste had been generated between 1950 and 2015 and only around 9% of this has been recycled, 12% was incinerated and 79% was accumulated in landfill and 3% in the ocean. Furthermore, Plastics Europe (2021) estimated global plastic production to be 367 million tonnes in 2020.

Geyer *et al.* (2017) and Hsu *et al.* (2022) are in agreement that the ensuing rapid growth in plastic production is extraordinary, surpassing most other man-made materials. So, if plastics are not properly managed, an epidemic may be inevitable as this is a collective failure of humankind (Wagh *et al.*, 2022). Additionally, the health challenges posed by plastic pollution indicate that there is an urgency to

DOI: 10.4324/9781003278443-15

tackle the challenge by preventing further disposal in landfills, drainages and oceans specifically in Africa (Kazançoğlu et al., 2021).

Africa is the world's second-largest continent after Asia (United Nations Environment Programme, 2018a). The plastic pollution challenge is exacerbated in Africa due to various factors such as infrastructure deficits (Oyinlola et al., 2018), growing population (Perkins, 2012), rural–urban migration (Yhdego, 1995), economic development (Rebellón, 2017) and cultural attitudes to waste (Adefila et al., 2020). According to United Nations Environment Programme (2018a), individual African countries are increasingly facing development challenges, one of which includes waste management and therefore by extension "plastic waste management" which is directly related to the rapid increase in Africa's population.

As countries in Africa continue to develop from low-income to middle-income economies, their waste management situation will continue to evolve (Perkins, 2012). Consequently, plastic consumption and pollution will continue to increase as these emerging economies grow. The emergence of plastic pollution has become a major driver for creating sustainable solutions to manage plastic waste challenges (Velis et al., 2021).

Plastic waste poses several challenges despite prevalent legal frameworks and policies regarding waste management (Thomas et al., 2022). This is increasing as Thomas et al. (2022) noted that the gap between legislature, waste management policy and actual waste management practices are widening. While Desmond and Asamba (2019) noted that the legal and regulatory framework to foster a green circular economy (CE) is still in its infancy in most African countries.

According to Khajuria et al. (2022), the CE approach and technological innovation have proven to be a highly efficient way to reduce final waste and decrease the use of natural resources, consequently identifying a circular approach as a viable intervention for managing plastic waste. The transition to a CE for plastics is high on the socioeconomic and political agendas of many countries (Maione et al., 2022). In the Global West, particularly Europe, there is a greater understanding of circular practices in multinational organisations with well-documented case studies and the emergence of new circular business models (Moreno et al., 2016). While in the Global South, particularly in Africa, there are small-scale examples of circular practices such as collection, reuse, repair and recycling which are gradually becoming recognised (Desmond & Asamba, 2019). Similarly, Desmond and Asamba (2019) stated that African case studies stay "hidden" as they are yet to be documented through academic research. However, several African countries are currently focused on developing green economies, and the effect is dependent on the CE strategy chosen by each country (Paiho et al., 2020). Also, these efforts can be optimised with the use of digital technology (DT) (Ellen Macarthur Foundation, 2019).

Data is a critical component for implementing systemic CE intervention for plastic pollution (Kolade et al., 2022; Oyinlola et al., 2022). In other words, verifiable data is required to develop the right tools for mapping and managing the entire plastic value chain (United Nations Environment Programme, 2018b). Plastic data within the value chain in the Global South is conflicting (Oyinlola et al., 2022), unverifiable (Williams-Wynn & Naidoo, 2020) and inadequate (Ryberg et al., 2019). Hence, it is quite difficult to truly understand the magnitude of the issue. This means stakeholders need to embrace DT solutions that will contribute towards data traceability and narrow the loop at the same time within the plastic value chain (Paiho et al., 2020).

This chapter contributes to the discourse on the global plastic crisis with particular emphasis on how plastic management in Africa can be enhanced with adequate data. This chapter highlights that effective data collection and usage will be facilitated by a multi-stakeholder, multi-process and multi-sectoral approach. This study, therefore, argues for a plastic data exchange (PDE) platform which will facilitate collaboration and exchange of data between stakeholders.

This chapter is organised as follows: First, it reviews the current state of the circular plastic economy (CPE) in Africa with a focus on data. This is followed by a discussion on the critical actors, elements, components, tools and technologies for an effective PDE platform for the plastic value chain in Africa. This chapter ends with a discussion of the interaction between the constituents of PDE, a recommendation for an effective PDE platform, and presents the conclusion from the study.

2   The CPE in Africa

As described in Section 1, plastic pollution is an urgent global challenge that most African countries are attempting to solve by adopting practices that will serve as a mechanism for moving from a linear economy to building a CE. According to Paiho et al. (2020), the challenges for policy, business, knowledge and technology must be addressed to build a CE. These challenges can be associated with the traditional, financial, political, social, institutional, regulatory and technical conditions of each country (Ghosh, 2019; Schroeder et al., 2023).

Policy instruments are generally the starting point to direct, monitor and report activities within the industry (Dagilienė et al., 2021). For example, in 2003, South African Government introduced a ban and levy on retailers' use of plastic bags; also in 2008, Rwanda Government introduced its ban on single-use plastics bags; and in 2016, the Mauritius Government banned the importation, manufacturing, sales and supply of plastic bags (United Nations Environment Programme, 2018a). However, despite most countries across the African continent having good policies, weak implementation remains a challenge for these policies to support a systemic change (Sylvester & Ikudayisi, 2021).

The transition within a business from a linear to a CE brings with it a range of practical challenges for companies (Bocken et al., 2016). These challenges make it difficult for many businesses to shift their operations rapidly. Nevertheless, businesses need to align for a wider impact in the CPE (Sukhdev *et al.*, n.d.). In addition, knowledge is a key challenge that must equally be addressed as it is important to change the mindset of people towards plastics and the CE ensuring that plastic waste is acknowledged as a resource that must be documented (Velis *et al.*, 2021).

Another critical requirement for the CE transition is the need for data across the entire value chain. Data availability is considered critical and essential for the plastic ecosystem as it will serve as an enabler for Africa's transition to a CPE (Kristoffersen *et al.*, 2020). Several scholars have highlighted the importance of data such as Ahmed *et al.* (2022) who suggested having several data-sharing platforms. Barrie *et al.* (2022) stated that data is important for bridging the circularity divide and suggested that valuable and verifiable data will enable stakeholders to operate efficiently. Hsu *et al.* (2022) noted that stakeholders need access to data, information and knowledge across the CE value chain to operate optimally.

However, access to data remains a big challenge in Africa which can be demonstrated by the inconsistency in published reports (Oyinlola *et al.*, 2022). For example, the quantity of mismanaged waste in Africa was reported by Jambeck *et al.* (2018) to be 4.4 million tonnes in 2010, while a report from United Nations Environment Programme (2018b) stated a much lower figure of 0.93 million tonnes in 2015 which implies about 80% reduction within 5 years. Similarly, Nwafor (2021) reported in an online article that Nigeria generates 32 million tonnes of waste of which 2.5 million tonnes are plastics, while a World Bank (2021) report noted that Nigeria generates 27.6 million tonnes of waste and 13% is plastics, while the Waste Management Society of Nigeria (WAMASON) estimates that Nigeria generates 65 million tonnes of waste per annum (All Africa, 2014). Most of these reports do not provide sufficient information to interrogate the robustness, accuracy and precision of their data. Therefore, it is presumed that plastic data collected is mostly based on broad estimations of various figures and statistics on plastic production, consumption, disposal and recycling. While this might be logical, Velis *et al.* (2021) noted, "There are bad waste data, worse waste data. Or … no waste data". This may indeed be the case for many African countries as plastic and plastic waste-related information is still largely unreliable (Ellen MacArthur Foundation, 2016), unpredictable (Pedro *et al.*, 2021) and sometimes unavailable (Soós *et al.*, 2021).

The preceding discussion highlights the need for a reliable PDE platform which will be an avenue for reporting, monitoring, accessing and verifying data within the plastic value chain. This must be robust, well organised, managed and governed within a structure to close the plastic waste data gap (de Sousa, 2021). An effective PDE platform will ultimately result in increasing CE practices and accelerate the application of technology in the ecosystem (Thomas *et al.*, 2022). Furthermore, the PDE platform will contribute to maintaining the environmental, social and governance (ESG) structure within the industry (Nyathi & Togo, 2020).

3   Elements of the PDE

This section draws on a practitioner's perspective to propose the critical elements for an effective PDE platform, specifically focusing on "Who", "What" and "How".

- "Who" refers to the organisations responsible for collecting, storing, processing and governing data collection.
- "What" refers to the kind and components of data that should be collected.
- "How" refers to the tools and technologies for data collection and dissemination.

3.1   Organisations and Institutions on the PDE Platform

Maione *et al.* (2022) noted that several scholars have recommended a shift from fragmented CE practices to a more systemic approach to closing the loop. In recent times, more collaborations and alignment have occurred across the circular plastic ecosystem (Mhatre *et al.*, 2021). As a result, information sharing is "seemingly" becoming better as roles are steadily being defined (Jabbour *et al.*, 2019). This collaboration is essential for an effective PDE platform. The organisations that should be within this PDE platform as well as their corresponding roles are discussed below.

3.1.1   Government

Governments have a critical role to play in the implementation of the PDE platform (Liu *et al.*, 2022) as it provides regulatory oversight functions across the value chain. The role of the government is vital in the process of collection, use and governance of data. Government agencies such as statistical bureaus, ministries of environment and associated parastatals have a responsibility to ensure a robust and continuous system for collecting data is established to facilitate the ease of data and information management (Maione *et al.*, 2022). In addition, the role of the government extends to monitoring and reporting within the PDE platform (Nyathi & Togo, 2020). Hence, a re-verification of the data collected is necessary. Secondly, data is a very essential information for government agencies as this helps to forecast and plan appropriate measures for the plastic value chain as well as contribute to efforts towards ensuring a sustainable smart city (Sukhdev *et al.*, n.d.). Thirdly, the government should be responsible for the governance of the PDE and consideration in the governance of data including data security (Sukhdev *et al.*, n.d.), funding (Dagilienė *et al.*, 2021), infrastructure and communication (Demestichas & Daskalakis, 2020). It is important to note that while the government may understand its role and see the benefit of the PDE platform, its efforts may be largely opportunistic, driven by bureaucracy or political agenda (Royle *et al.*, 2022).

3.1.2 Producer Responsibility Organisations (PROs)

PROs support the implementation of the extended producer responsibility (EPR) policy. The EPR is an increasingly popular policy instrument that mandates producers of plastics have a responsibility in post-consumer recovery to reduce toxicity and waste (Perkins, 2012). One example of a PRO is Food and Beverage Recycling Alliance (FBRA) in Nigeria which ensures (1) better product design, (2) drives collection at scale to create demand for recyclables, (3) increases recycled content, (4) fosters partnerships and collaboration among stakeholders (FBRA, 2018) and (5) provides financial and operational support for business development (United Nations Environment Programme, 2018a). Another example is the "African Plastics Recycling Alliance" which was established in 2019 by multinationals such as Nestle, Coca-Cola and Pepsi in a bid to address the end of life of the plastic value chain (Break Free From Plastic, 2019). This alliance aims to improve the plastic recycling infrastructure across sub-Saharan Africa (IISD, 2019). These organisations need data to operate, and they can also contribute valuable data to the PDE platform.

3.1.3 Producers

These are organisations that manufacture plastics or products required for plastic packaging, automotive, electronics, building and construction (Sylvester & Ikudayisi, 2021). Their role will be to provide accurate data on production volume and plastic type and composition. Within the EPR, they are primarily responsible for the end-of-life management of their product (Ellen MacArthur Foundation, 2019). Producers will contribute to the PDE platform by regularly supplying accurate and verifiable data for products produced and recovered. This will reduce false data on plastic production significantly and will help the ecosystem towards recovery efforts.

3.1.4 Collection Companies

This generally has three categories. The "formal" and "informal" collectors are the common name for those operating in the collection space (Ellen MacArthur Foundation, 2014; Thomas *et al.*, 2022). The formal collectors are known to operate a structured and integrated method of collection (Ellen MacArthur Foundation, 2014), while the informal sector still operates a very casual and unplanned collection method (Pedro *et al.*, 2021). The effectiveness of their collection system is still debated; however, their role is vital to the collection ecosystem (Ellen MacArthur Foundation, 2019). In addition, the third category is "semi-formal" collectors which must be factored in. This collection style is regulated by government collection services which are often regarded as a public–private partnership (PPP) (Perkins, 2012). Data generated from these three entities is done manually and remains fragmented, but technology can play a significant role in ensuring information is collected (Kamble *et al.*, 2021) and accurately analysed (Halog & Anieke, 2021).

3.1.5 Non-Governmental Organisations (NGOs)

Within the plastic economy ecosystem, a few entities have been identified as NGOs. The role of the NGOs will be significant, especially from the local income areas' (grassroots) perspective and particularly when it comes to women and youth (Khajuria *et al.*, 2022). The role of the NGOs is not limited as they operate a flexible system that can be adapted within the PDE platform. Their role can be extended to capturing communication data, campaigns and initiatives data and knowledge transfer data at local and state levels that will be relevant to the sector. For example, associations such as the Lagos State Recyclers Association of Nigeria (LAGRA), the Circular Economy Innovation Partnership (CEIP, 2022) in Nigeria, the Africa Circular Economy Network (ACEN, 2022) in South Africa, the South African Plastic Recycling Organisation (Sapro, 2019) have existing structures that can be plugged into the data collection network

3.1.6 Academia

Academia will play a role in finding pathways for the PDE platform stakeholders to connect effectively. Their roles will also extend to data analysis from various sources such as primary and secondary sources. Also, the CE is becoming a very popular subject area within academia. They will be effective in offering recommendations, developing frameworks, mapping out interactions and modelling solutions to all stakeholders (Demestichas & Daskalakis, 2020).

3.1.7 Technology Companies

Several companies across the continent are harnessing or developing technology for the CPE. These entities are regarded as the newest entry within the ecosystem, although experiencing slow acceptance which can pose a challenge for data collection (Ellen Macarthur Foundation, 2019). However, the role of technology is important as it can effectively, efficiently and remotely collect data on the performance and usage of products (Alcayaga *et al.*, 2019). The introduction of technology shows a promising approach to bringing all stakeholders together to develop the PDE platform. Hence, it presents a huge opportunity for the PDE concept and assists in clearly defining roles in the PDE platform. For instance, Pakam Technology, a marketplace for waste recycling in Nigeria, has positioned itself as a household name in the recycling industry by connecting waste generators to waste collectors in real time (Pakam, 2021). Another technology in Uganda, Yo-Waste, aims to build a zero-waste community (yo-waste, 2018). In both cases, data is collected directly from consumers, which can help to understand consumer behaviour.

TABLE 12.1 Component of data collection

	Social	*Technical*	*Governance*
Production	Producers Job opportunities Location	Manufacturing Process Type of plastic Material properties	Traceability Quantity EPR
Distribution	Retailers Wholesalers Job opportunities	Logistics	Traceability Export/import policies

Use	Types of consumers
	Gender
	Level of education
	Level of awareness
	Consumption habits/
	  behaviour
		Unstructured
		  supplementary
		  service data (USSD)
		App
		Instant messaging
			Tracking
			Consumption
			  bill

Collection	Informal/semi-formal/
	  formal collectors
	Job opportunities for
	  women/youths (gender)
	Incentives
		Collection
		  infrastructure
		Logistics
			Trading

Recycling	Pricing	Re-manufacturing Recycling process Material recovery facilities Reverse logistics	Trading Standardisation Pricing

3.2 Components of Data on the PDE Platform

According to Maione *et al.* (2022), information factors are often absent or merely discussed which fails to capture the importance of data traceability and proper reporting on plastic materials and pollution along the value chain. Due to the fragmented approach, limited data is available (Maione *et al.*, 2022). Table 12.1 proposes the basic components that each stage of the value chain should have at the very least. These components will lead to the fit-for-purpose design and quick successful implementation of CE solutions. Table 12.1 shows that the PDE will rely on a mixture of quantitative and qualitative methods at various stages of the plastic value chain.

3.3 Tools and Technologies for the PDE

As the usage of the Internet of things (IoT) grows, so does the advancement in the use of technology applications for solving real-life problems (Alcayaga *et al.*, 2019). Africa has come a long way with technology penetration, especially with the emergence of mobile phones, an increase in Internet penetration (Giuliani &

Ajadi, 2019) and tech hubs (Atiase *et al.*, 2020). Although the boom in smartphone penetration across the continent (GSMA, 2020) has made information exchange easier, as it has provided the opportunity to capture various forms of data such as pictures, and transfer information via instant messaging such as WhatsApp, it is yet to solve the challenge of "accurate and verifiable data".

Technologies such as blockchain and artificial intelligence (AI) can speed up the PDE process in collecting and analysing data (Tseng *et al.*, 2018). According to Sukhdev *et al.* (n.d.), DT can ensure (1) asset tagging, (2) geo-spatial information, (3) big data management and (4) connectivity. For example, using AI will enhance the development of new products, combine real-time and historical data of products or services and optimise circular infrastructure (Ellen Macarthur Foundation, 2019). AI has the potential to provide systematic development algorithms that will collect and analyse data which will prevent the falsification of numbers and information. In other words, there needs to be a remarkably simple, yet sophisticated and affordable digital process for optimising the operations of Table 12.1 in the PDE. Technology is needed to enable the communication within the PDE ecosystem (Demestichas & Daskalakis, 2020).

4 Discussion and Recommendation

4.1 Discussion

The PDE platform will leverage DT and CE, which are both considered to be emerging fields (Kristoffersen *et al.*, 2020). The Ellen MacArthur Foundation (2019) noted that if combined, both principles have the potential to support and enable a systemic shift. The PDE platform provides a pathway for capturing essential data within the plastic value chain and eliminates the bias of working in silos, fear of idea duplication, stealing and double taxation among others, from stakeholders. This presents an opportunity for the CPE to establish sustainable development goal (SDG) 17 "partnership for the goal" as a core goal through a data-sharing platform. The data-sharing platform will serve as a central warehousing system that can be executed through a Public Private Partnership (PPP) arrangement managed by the government and private sector. The system will connect plastic in production (PIP), plastic in transit (PIT), plastic in circulation (PIC) and plastic in recycling (PIR) through data collected from Table 12.1. Having plastics properly categorised into the right data set will ensure plastic materials fall under the right group which will benefit all stakeholders for resource efficiency. The PDE platform described in this chapter focused only on "Who", "What" and "How". However, adopting the PDE is inherently complex, and systems thinking is essential for wider impact (Bocken *et al.*, 2016). Therefore, a conceptual framework is fundamental to progressing the PDE concept further. This framework should explore the stakeholders, key factors, variables and constraints and show their interrelationships (Alcayaga *et al.*, 2019).

It is almost impossible to develop a framework without some form of data and impossible to gather data without some form of framework which shows they are dependent variables. A few frameworks have been proposed for the CE (Lasi *et al.*, 2014). For example, "the circular strategies scanner" provides comprehensive support for plastic manufacturing companies (Charnley *et al.*, 2019). Bocken *et al.* (2016) proposed a framework for product design and business models within the CE. Royle *et al.* (2022) proposed the plastic drawdown (PD) framework, a boundary-spanning tool between plastic pollution knowledge and policy design. However, the existing frameworks do not fully capture how data can be collected and analysed for a CPE. According to Charnley *et al.* (2019), data-driven intelligence is gradually becoming a pervasive feature of our economy. Hence, there is the need for creating a data-driven framework that will enable the activities within the PDE platform. In addition, plastic awareness, communication and educational (PACE) programmes must be included to enhance plastic data management across the continent (Sylvester & Ikudayisi, 2021). However, enforcement may remain a major problem. Therefore, a well-captured plastic data framework can present an opportunity to further expose the lapses within the plastic industry.

4.2 Recommendations

Given the above-highlighted problems, associated with data capture in Africa, the following recommendations are made:

1. A data exchange platform should be developed as a framework to set the tools and parameters to capture data in the CPE.
2. All roles within the data exchange platform must be clearly defined with clear input and output expectations.
3. Ensuring a closed-loop data collection process, especially for the informal, semi-formal and formal categories.
4. Public communication and awareness campaigns should expressly carry the need for and importance of data.
5. Verification rewards can be introduced at all levels of participation to prevent ambiguity.
6. A marketplace can be created for data exchange across countries.
7. Enforcement of EPR policy is still a starting point for the platform; the government should decentralise information from the federal and state level to the local government level. This will ensure enforcement and promote a resource-efficient economy.

5 Conclusion

Many African countries have witnessed this rapid growth in the last few decades, especially in population which has also increased pollution, and unfortunately, no blueprint was prepared for this plastic pollution epidemic. Therefore, a systematic study must be done to answer related questions about plastic waste pollution and plastic management. It's impossible to control what we cannot measure as most countries operate a reactive system instead of a proactive system.

Plastic material and plastic waste are considered a resource within the CE. Therefore, it is important to treat them as such. However, a proper data system to close the loop effectively will contribute to resolving non-statistical data on plastic materials in most developing countries.

This chapter has proposed a PDE platform, which has the potential to improve transparency and ensure standardised and streamlined processes. In addition, ensure centralised collection and reporting that can transform data into insights and trends. However, this does not come without its limitations, which need to be further explored within the context of a framework. It highlights that effective data collection and usage will be accelerated by a PDE platform which will facilitate collaboration between stakeholders.

References

ACEN. (2022). *African Circular Economy* Network website, accessed 29 November 2022, www.acen.africa.

Adefila, A., Abuzeinab, A., Whitehead, T., & Oyinlola, M. (2020). Bottle house: utilising appreciative inquiry to develop a user acceptance model. *Built Environment Project and Asset Management, 10*(4), 567–583. https://doi.org/10.1108/BEPAM-08-2019-0072

Ahmed, A. A., Nazzal, M. A., Darras, B. M., & Deiab, I. M. (2022). A comprehensive multi-level circular economy assessment framework. *Sustainable Production and Consumption, 32*, 700–717. https://doi.org/10.1016/J.SPC.2022.05.025

Alcayaga, A., Wiener, M., & Hansen, E. G. (2019). Towards a framework of smart-circular systems: an integrative literature review. *Journal of Cleaner Production, 221*, 622–634. https://doi.org/10.1016/j.jclepro.2019.02.085

All Africa. (2014). AllAfrica-All the Time, accessed 28 November 2022, https://allafrica.com/stories/201406061005.html

Atiase, V. Y., Kolade, O., & Liedong, T. A. (2020). The emergence and strategy of tech hubs in Africa: implications for knowledge production and value creation. *Technological Forecasting and Social Change, 161*, 120307. https://doi.org/10.1016/J.TECHFORE.2020.120307

Barrie, J., Anantharaman, M., Oyinlola, M., & Schröder, P. (2022). The circularity divide: what is it? And how do we avoid it? *Resources, Conservation and Recycling, 180*, 106208. https://doi.org/10.1016/J.RESCONREC.2022.106208

Bocken, N. M. P., de Pauw, I., Bakker, C., & van der Grinten, B. (2016). Product design and business model strategies for a circular economy. *Journal of Industrial and Production Engineering, 33*(5), 308–320. https://doi.org/10.1080/21681015.2016.1172124

Break Free From Plastic. (2019). Branded Vol. II: Identifying the World's top corporate plastic polluters, II, 77. Retrieved from www.breakfreefromplastic.org/wp-content/uploads/2020/07/branded-2019.pdf.

CEIP. (2022). *Circular Economy Innovation Partnership webiste*, accessed 29 November 2022, https://ceipafrica.org

Charnley, F., Tiwari, D., Hutabarat, W., Moreno, M., Okorie, O., & Tiwari, A. (2019). Simulation to enable a data-driven circular economy. *Sustainability, 11*(12). https://doi.org/10.3390/su11123379

Dagilienė, L., Varaniūtė, V., & Bruneckienė, J. (2021). Local governments' perspective on implementing the circular economy: a framework for future solutions. *Journal of Cleaner Production, 310*, 127340. https://doi.org/10.1016/J.JCLEPRO.2021.127340

Dasgupta, S., Sarraf, M., & Wheeler, D. (2022). Plastic waste cleanup priorities to reduce marine pollution: a spatiotemporal analysis for Accra and Lagos with satellite data. *Science of the Total Environment, 839*(March), 156319. https://doi.org/10.1016/j.scitotenv.2022.156319

Demestichas, K., & Daskalakis, E. (2020). Information and communication technology solutions for the circular economy. *Sustainability, 12*(18). https://doi.org/10.3390/su12187272

Desmond, P., & Asamba, M. (2019). Accelerating the transition to a circular economy in Africa: case studies from Kenya and South Africa. *The Circular Economy and the Global South*, 152–172. https://doi.org/10.4324/9780429434006-9

de Sousa, F. D. B. (2021). Management of plastic waste: a bibliometric mapping and analysis. *Waste Management and Research, 39*(5). https://doi.org/10.1177/0734242X21992422

Ellen MacArthur Foundation. (2014). *Towards the Circular Economy*. 113–123. www.mckinsey.com/~/media/mckinsey/dotcom/client_service/sustainability/pdfs/towards_the_circular_economy.ashx

Ellen MacArthur Foundation. (2016). Circular economy in India: rethinking growth for long-term prosperity (executive summary). *Ellen MacArthur Foundation*, 1–86.

Ellen MacArthur Foundation. (2019). *Artificial Intelligence and the Circular Economy Ai As a Tool to Accelerate*. September, 1–62.

Ellen MacArthur Foundation, United Nations Environment Programme, & New plastic Economy. (2019). *The New Plastics Economy Global Commitment 2019 Progress Report* (Issue October), 6–51. https://www.ellenmacarthurfoundation.org/news/first-annual-new-plastics-economy-global-commitment-progress-report-published

FBRA. (2018). *The Food & Beverage Recycling Alliance*. Accessed 29 November 2022, www.fbranigeria.ng

Geyer, R., Jambeck, J. R., & Law, K. L. (2017). Production, use, and fate of all plastics ever made. *Science Advances, 3*(7), e1700782. https://doi.org/10.1126/sciadv.1700782

Ghosh, S. K. (2019). Circular economy: global perspective. *Circular Economy: Global Perspective*. https://doi.org/10.1007/978-981-15-1052-6

Giuliani, D., & Ajadi, S. (2019). *618 Active Tech Hubs: The Backbone of AFRICA'S Tech Ecosystem*. Mobile for Development. www.gsma.com/mobilefordevelopment/blog/618-active-tech-hubs-the-backbone-of-africas-tech-ecosystem/

GSMA. (2020). *The Mobile Economy*. www.gsma.com/mobileeconomy/#

Halog, A., & Anieke, S. (2021). A review of circular economy studies in developed countries and its potential adoption in developing countries. *Circular Economy and Sustainability, 1*(1), 209–230. https://doi.org/10.1007/s43615-021-00017-0

Hsu, W. T., Domenech, T., & McDowall, W. (2022). Closing the loop on plastics in Europe: the role of data, information and knowledge. *Sustainable Production and Consumption, 33*, 942–951. https://doi.org/10.1016/J.SPC.2022.08.019

IISD. (2019). *Companies Launch African Plastics Recycling Alliance*. SDG KNOWLEDGE HUB. https://sdg.iisd.org/news/companies-launch-african-plastics-recycling-alliance/

Jabbour, C. J. C., de S. Jabbour, A. B. L., Sarkis, J., & Filho, M. G. (2019). Unlocking the circular economy through new business models based on large-scale data: an integrative framework and research agenda. *Technological Forecasting and Social Change, 144*, 546–552. https://doi.org/10.1016/J.TECHFORE.2017.09.010

Jambeck, J., Hardesty, B. D., Brooks, A. L., Friend, T., Teleki, K., Fabres, J., Beaudoin, Y., Bamba, A., Francis, J., Ribbink, A. J., Baleta, T., Bouwman, H., Knox, J., & Wilcox, C. (2018). Challenges and emerging solutions to the land-based plastic waste issue in Africa. *Marine Policy, 96*, 256–263. https://doi.org/10.1016/j.marpol.2017.10.041

Kamble, S. S., Belhadi, A., Gunasekaran, A., Ganapathy, L., & Verma, S. (2021). A large multi-group decision-making technique for prioritizing the big data-driven circular economy practices in the automobile component manufacturing industry. *Technological Forecasting and Social Change, 165*, 120567. https://doi.org/10.1016/J.TECHFORE.2020.120567

Kazançoğlu, Y., Sağnak, M., Lafcı, Ç., Luthra, S., Kumar, A., & Taçoğlu, C. (2021). Big data-enabled solutions framework to overcoming the barriers to circular economy initiatives in healthcare sector. *International Journal of Environmental Research and Public Health, 18*(14). https://doi.org/10.3390/ijerph18147513

Khajuria, A., Atienza, V. A., Chavanich, S., Henning, W., Islam, I., Kral, U., Liu, M., Liu, X., Murthy, I. K., Oyedotun, T. D. T., Verma, P., Xu, G., Zeng, X., & Li, J. (2022). Accelerating circular economy solutions to achieve the 2030 agenda for sustainable development goals. *Circular Economy, 1*(1), 100001. https://doi.org/10.1016/j.cec.2022.100001

Kolade, O., Odumuyiwa, V., Abolfathi, S., Schröder, P., Wakunuma, K., Akanmu, I., Whitehead, T., Tijani, B., & Oyinlola, M. (2022). Technology acceptance and readiness of stakeholders for transitioning to a circular plastic economy in Africa. *Technological Forecasting and Social Change, 183*, 121954. https://doi.org/10.1016/J.TECHFORE.2022.121954

Kristoffersen, E., Blomsma, F., Mikalef, P., & Li, J. (2020). The smart circular economy: a digital-enabled circular strategies framework for manufacturing companies. *Journal of Business Research, 120*, 241–261. https://doi.org/10.1016/J.JBUSRES.2020.07.044

Lasi, H., Fettke, P., Kemper, H.-G., Feld, T., & Hoffmann, M. (2014). Industry 4.0. *Business & Information Systems Engineering, 6*(4), 239–242.

Liu, Q., Trevisan, A. H., Yang, M., & Mascarenhas, J. (2022). A framework of digital technologies for the circular economy: digital functions and mechanisms. *Business Strategy and the Environment, 31*(5), 2171–2192. https://doi.org/10.1002/bse.3015

Lopez-Aguilar, J. F., Sevigné-Itoiz, E., Maspoch, M. L., & Peña, J. (2022). A realistic material flow analysis for end-of-life plastic packaging management in Spain: data gaps and suggestions for improvements towards effective recyclability. *Sustainable Production and Consumption, 31*. https://doi.org/10.1016/j.spc.2022.02.011

Maione, C., Lapko, Y., & Trucco, P. (2022). Towards a circular economy for the plastic packaging sector: insights from the Italian case. *Sustainable Production and Consumption, 34*, 78–89. https://doi.org/10.1016/j.spc.2022.09.002

Mhatre, P., Panchal, R., Singh, A., & Bibyan, S. (2021). A systematic literature review on the circular economy initiatives in the European Union. *Sustainable Production and Consumption, 26*, 187–202. https://doi.org/10.1016/J.SPC.2020.09.008

Moreno, M., De los Rios, C., Rowe, Z., & Charnley, F. (2016). A conceptual framework for circular design. *Sustainability, 8*(9). 937. https://doi.org/10.3390/su8090937

Nwafor, J. (2021). Fighting plastic waste: a double-edged sword. *SciDev. Net.* Availaible at https://www.scidev.net/sub-saharan-africa/multimedia/fighting-plastic-waste-a-dou ble-edged-sword/, accessed 23 March 2023.

Nyathi, B., & Togo, C. A. (2020). Overview of legal and policy framework approaches for plastic bag waste management in African countries. *Journal of Environmental and Public Health.* https://doi.org/10.1155/2020/8892773

Oyinlola, M., Schröder, P., Whitehead, T., Kolade, S., Wakunuma, K., Sharifi, S., Rawn, B., Odumuyiwa, V., Lendelvo, S., Brighty, G., Tijani, B., Jaiyeola, T., Lindunda, L., Mtonga, R., & Abolfathi, S. (2022). Digital innovations for transitioning to circular plastic value chains in Africa. *Africa Journal of Management, 8*(1), 83–108. https://doi. org/10.1080/23322373.2021.1999750

Oyinlola, M., Whitehead, T., Abuzeinab, A., Adefila, A., Akinola, Y., Anafi, F., Farukh, F., Jegede, O., Kandan, K., Kim, B., & Mosugu, E. (2018). Bottle house: a case study of transdisciplinary research for tackling global challenges. *Habitat International, 79,* 18–29. https://doi.org/10.1016/j.habitatint.2018.07.007

Paiho, S., Mäki, E., Wessberg, N., Paavola, M., Tuominen, P., Antikainen, M., Heikkilä, J., Rozado, C. A., & Jung, N. (2020). Towards circular cities—conceptualizing core aspects. *Sustainable Cities and Society, 59*(April), 102143. https://doi.org/10.1016/ j.scs.2020.102143

Pakam. (2021). *Pakam Technology,* accessed 29 November 2022,https://www.pakam.ng.

Pedro, F., Giglio, E., Velazquez, L., & Munguia, N. (2021). Constructed governance as solution to conflicts in E-waste recycling networks. *Sustainability.* https://doi.org/ 10.3390/su13041701

Perkins, S. (2012). What a waste. *Drug Topics, 156*(2), 296.

Plastics Europe. (2021). Plastics the fact 2021. In *Plastics Europe Market Research Group (PEMRG) and Conversio Market & Strategy GmbH,* 1–34. https://plasticseurope.org/

Rebellón, L. (2017). Waste management waste management. *Group, 7*(888), 195–200. www.scopus.com/inward/record.uri?eid=2-s2.0-0022180155&partnerID=40&md5= 18a8ce1aa279d51fd15e204b9e4623d0

Royle, J., Jack, B., Parris, H., Elliott, T., Castillo, A. C., Kalawana, S., Nashfa, H., & Woodall, L. C. (2022). Plastic drawdown: A rapid assessment tool for developing national responses to plastic pollution when data availability is limited, as demonstrated in the Maldives. *Global Environmental Change, 72,* 102442. https://doi.org/10.1016/ j.gloenvcha.2021.102442

Ryberg, M. W., Hauschild, M. Z., Wang, F., Averous-Monnery, S., & Laurent, A. (2019). Global environmental losses of plastics across their value chains. *Resources, Conservation and Recycling, 151,* 104459.

Sapro. (2019). *South African Plastic Recycling Organisation.* www.plasticrecyclingsa.co.za/, accessed 29 November 2022.

Schroeder, P., Oyinlola, M., Barrie, J., Bonmwa, F., & Abolfathi, S. (2023). Making policy work for Africa's circular plastics economy. *Resources, Conservation & Recycling, 190,* 106868. https://doi.org/10.1016/j.resconrec.2023.106868

Soós, R., Whiteman, A., Gavgas, G., Charnley, F., Tiwari, D., Hutabarat, W., Moreno, M., Okorie, O., Tiwari, A., de Sousa, F. D. B., Satispi, E., & Samudra, A. A. (2021). The cost of preventing ocean plastic pollution. *Waste Management and Research, 39*(5), 664–678. https://doi.org/10.1787/5c41963b-en

Sukhdev, A., Vol, J., Brandt, K., & Yeoman, R. (2018). Cities in the circular economy: the role of digital technology. Ellen MacArthur Foundation: Cowes, UK, https://emf.thi rdlight.com/link/41iwzsqtagzz-dhmjn3/@/preview/1?o

Sylvester, O., & Ikudayisi, O. (2021). An overview of solid waste in Nigeria: challenges and management. *Jordan Journal of Earth and Environmental Sciences, 12*(1), 36–43.

Thomas, O., Šebo, J., Šebová, M., Palčič, I., Hopewell, J., Dvorak, R., Kosior, E., Mwanza, B. G., Mbohwa, C., Telukdarie, A., Allen-Taylor, K. O., Kolade, O., Odumuyiwa, V., Abolfathi, S., Schröder, P., Wakunuma, K., Akanmu, I., Whitehead, T., Tijani, B., … Li, J. (2022). Strategies for the recovery and recycling of plastic solid waste (PSW): a focus on plastic manufacturing companies. *Journal of Global Ecology and Environment, 16*(August), 1–10. https://doi.org/10.1016/j.techfore.2022.121954

Tseng, M. L., Tan, R. R., Chiu, A. S. F., Chien, C. F., & Kuo, T. C. (2018). Circular economy meets industry 4.0: can big data drive industrial symbiosis? *Resources, Conservation and Recycling, 131*, 146–147. https://doi.org/10.1016/J.RESCONREC.2017.12.028

UNEP. (2018a). *Africa Waste Management Outlook.* www.unep.org/ietc/resources/publicat ion/africa-waste-management-outlook

UNEP. (2018b). *Mapping of Global Plastics Value Chain and Plastics Losses to the Environment: With a Particular Focus on Marine Environment.* United Nations Environment Programme. http://hdl.handle.net/20.500.11822/26745

Velis, C. A., Cook, E., & Cottom, J. (2021). Waste management needs a data revolution – Is plastic pollution an opportunity? *Waste Management and Research, 39*(9), 1113–1115. https://doi.org/10.1177/0734242X211051199

Wagh, A., Bhavsar, F., & Shinde, R. (2022). Plastic pollution. *International Journal for Research in Applied Science and Engineering Technology, 10*(3). https://doi.org/10.22214/ijraset.2022.40921

Williams-Wynn, M. D., & Naidoo, P. (2020). A review of the treatment options for marine plastic waste in South Africa. *Marine Pollution Bulletin, 161*, 111785. https://doi.org/10.1016/j.marpolbul.2020.111785

World Bank. (2021). *Skills Development.* www.worldbank.org/en/topic/skillsdevelopment

Yhdego, M. (1995). Urban solid waste management in Tanzania Issues, concepts and challenges. *Resources, Conservation and Recycling, 14*(1), 1–10.

yo-waste. (2018). *Yo-Waste,* accessed 29 November 2022, https://yowasteapp.com/

13

ENHANCING DECENTRALISED RECYCLING SOLUTIONS WITH DIGITAL TECHNOLOGIES

Silifat Abimbola Okoya, Muyiwa Oyinlola, Patrick Schröder, Oluwaseun Kolade and Soroush Abolfathi

1 Introduction

Plastics are a very valuable commodity due to their attractive material properties such as strength, flexibility and light weight which makes them ideal for a range of applications including medical services, building, transportation and most importantly packaging (Baran, 2020; Narancic and O'Connor, 2019). However, poor plastic waste management and lack of robust recycling mechanisms have transformed plastics into one of the most significant threats to the environment, accounting for around 10% of general waste (Barnes et al., 2009; Hopewell et al., 2009; OECD, 2022). The use of plastics as packaging materials, which represents over a third of plastics produced, has precipitated a global expansion of plastic production. Consumers have increasingly embraced a throwaway culture by moving from reusable to single-use containers that are disposed within a year, thereby exacerbating the challenge of municipal solid waste (Jambeck et al., 2015). Discarding of packaging materials has been reported to have a substantial impact on the environment especially because a high proportion of these are shipped to landfill within a year of production (Barnes et al., 2009; Hopewell et al., 2009). This has led to what some have coined the "Plastic Age" (Thompson et al., 2009). Today, the threat has evolved into a major crisis with significant contributions to climate change and greenhouse emissions (Shen et al., 2020).

Global production of plastics has steadily increased over the past 50 years. For example, in 2016, global production of plastic was 335 million tonnes per annum (Drzyzga and Prieto, 2019). If the current trajectory continues, it is estimated to grow to 33 billion metric tonnes by 2050 (Jambeck et al., 2015; Rochman et al., 2013). This trend has become a major concern as plastic is non-biodegradable, and microplastics are permeating into the food chain and atmosphere (Wright

DOI: 10.4324/9781003278443-16

and Kelly, 2017). Nonetheless, plastics remain essential products in the modern era largely due to the advancements in information technology, intelligent and smart packaging systems (Singh and Sharma, 2016). The main challenge is that plastic management has developed into a worldwide crisis as production has increased by 122-fold in 40 years (Joshi et al., 2019) with roughly 12.2 billion metric tonnes discarded as waste annually. That is approximately 3.9 billion metric tonnes of waste mishandled on land and 1.6 billion polluting the oceans (Jambeck et al., 2015; Rochman et al., 2013), especially the aquatic ecosystem (Bläsing and Amelung, 2018). The challenge is so pervasive that plastic waste has been proposed as a geological gauge for the Anthropocene era (Waters et al., 2016). Therefore, plastic has now emerged as a new planetary boundary menace (Galloway and Lewis, 2016; Rockström et al., 2009).

On average, high-income countries (HICs) create more plastic wastes per individual (Ritchie and Roser, 2018) as global solid waste generation correlates with gross national income per capita (Hoornweg et al., 2013; Wilson et al., 2015). In these countries, the deployment of advanced centralised recycling facilities makes the management of plastic waste more efficient. However, in Africa, the plastic pollution challenge is exacerbated by a lack of robust infrastructure and waste management systems (Oyinlola et al., 2018). The United Nations Environment Programme (UNEP) estimated that in 2015, Africa accounted for 24% of the world's total mismanaged plastic waste (UNEP, 2018a). The primary origins of waste in Africa are from households, open markets, formal institutions, public and commercial areas and the manufacturing companies (Kaseva and Mbuligwe, 2005). A United Nations (UN) report on municipal solid waste estimates that 99% of items purchased annually by consumers would be converted to waste approximately within the first six months (Ayeleru et al., 2020), with plastic being a significant player (Wilcox et al., 2015). Furthermore, fast-moving consumer goods have been found to dominate the plastic waste stream. For example, drinking water supply chain and sachet water have been identified as the biggest contributors to plastic waste in Africa. These have contributed to other issues such as clogged drains, breeding mosquitoes and localising floods (Williams et al., 2019).

The challenge of plastic waste management in Africa is exacerbated by growing population, varying consumer trends and increased urbanisation. For example, the complications of solid waste are further heightened by the increased rural to urban migration, bringing additional pressures on already overextended resources in the big cities (Yhdego, 1995). An important consideration in Africa is control of the entire value chain. Plastics enter the ecosystem from various entry points (Geyer et al., 2017), and waste pickers are not enough to meet the need. This is particularly difficult due to the lack of resources required to address issues associated with distance to disposal locations, street dumping and indiscriminate waste burning (Joshi et al., 2019). Scholars have compared the contemporary situation in Africa with the 1950s and 1960s, where waste management was efficient due to lower urban population and relatively adequate infrastructure (Achankeng, 2003; Adedibu and Okekunle, 1989; Henry et al., 2006; Kaseva and Mbuligwe, 2005).

In recent times, it was reported that the Covid-19 pandemic resulted in a change in consumer behaviour (Vanapalli et al., 2021). The situation was further aggravated by the temporary easing of the bans on single-use plastics which might have future consequences on transitioning to a circular economy (Vanapalli et al., 2021).

Plastic recycling is recognised as the most advanced method of sustainably dealing with the plastic pollution challenge (Hopewell et al., 2009; Zhong and Pearce, 2018). Prior to 1980, plastic recycling was insignificant, with the exception of non-fibre plastics (Geyer et al., 2017). While recycling rates have grown over the past decades, this growth has not been uniform across the world, a situation that has been described as the circularity divide (Barrie et al., 2022). For example, as of 2018, 32.5% of the 61.8 million tonnes of plastic produced in Europe was recycled (Plastics Europe, 2018). This compared to less than 10% in Africa (UNEP, 2018b). Despite this increase in recycling rates, the quantity of recycled plastics remains very low, and there has been continuous efforts by governments and municipalities across the world to push positive consumer recycling behaviour. Hornik et al. (1995) categorised consumer recycling behaviour into four generic groups: Firstly, intrinsic incentives such as interest in recycling determine consumers' attitudes to recycling. Secondly, extrinsic incentives such as rewards consumers receive for participating in recycling schemes are important. Thirdly, they noted that internal facilitators such as knowledge of the importance of recycling as well as the awareness of recycling programmes drive consumer behaviour. Fourthly, external facilitators such as convenience of participating in recycling programmes will usually trump incentives (Hornik et al., 1995). The gross domestic product (GDP) level and waste separation structure are critical factors that impact waste sorting, at a high or low rate, by residents based on different considerations when considering recycling behaviours as a high rate is mainly achieved when socio-demographic and external conditions are uniquely combined (Wan and Wan, 2020).

The earlier concept of recycling, especially within the industrial communities, was originally likened to the sphere of morality with the belief in right or wrong (Thøgersen, 1996). Several explanations have been postulated for the low recycling rates in low- and middle-income countries; for example, Kolade et al. (2022b) suggested that environmental concerns are usually not a priority as the majority of the population is still struggling to meet the necessities of life, such as food and shelter. Furthermore, it has been widely reported that plastic recycling is not always economically viable especially in Africa (Kreiger et al., 2014; Santander et al., 2020) as the costs of virgin plastics are usually cheaper than recycled plastics. Multinationals across Africa, such as Nestle, Coca-Cola and Pepsi, launched the "African Plastics Recycling Alliance" in 2019 (Break Free From Plastic, 2019). These corporations are among the biggest producers and distributors of fast-moving consumer goods with plastic packaging. This alliance aimed to address the end of life of the plastic value chain by improving the plastics recycling infrastructure across sub-Saharan Africa (IISD, 2019).

2 Plastic Recycling in Africa

In the Global West, recycling is usually done through centralised networks which leverage economies of scale associated with recycling of low-value products (Kreiger et al., 2014; Santander et al., 2020). This centralised approach involves the cost of transportation of high-volume and low-weight polymers (Kreiger et al., 2014; Santander et al., 2020). It can have significant environmental pollution impacts (Ragaert et al., 2017) due to the greenhouse gas emissions associated with collection and transportation (Garmulewicz et al., 2016). Although there is limited research on the comparative merits of centralised, clustered and decentralised waste management technologies, Anwar et al. (2018) noted that centralised systems realise more net profit in comparison to the decentralised and clustered methods. In practice, the approach adopted varies based on available infrastructure (Oyinlola et al., 2023b). On a large scale, plastic waste management necessitates the development of technology infrastructures. This is required at both the national and local levels in Africa guided by economic and political ability (Schroeder et al., 2023; Wilson et al., 2013). Due to resource constraints, there is a growing interest in locally managed decentralised circular economy (LMDCE) (Joshi and Seay, 2020), to allow sustainable recycling solutions which enable communities to take ownership and control of their waste. These decentralised solutions are driven by Industry 4.0 technologies such as three-dimensional (3D) printing, which upends the economies of scale, in favour of the economy of one associated with production of customised products using plastic waste as raw materials (Kolade et al., 2022a). In developing local technologies for decentralised plastic recycling, considerations need to be in place to ensure the suggested solutions are low cost, economically feasible, environment-friendly and socially suitable for it to be successful in Africa (Joshi et al., 2019). An example is the use of locally found plastic waste which was converted to plastic-derived fuel oil (PDFO) in Uganda (Joshi et al., 2019; Joshi and Seay, 2020; Schumacher, 2011). It is limited to polyolefin plastics, and although not a global solution to the plastic challenge, it is an example of a low-cost solution that can be implemented with a finite technology infrastructure (Browning et al., 2021).

Currently, there are only 67 plastic recycling plants registered in the African plastic recycling plant directory, with South Africa (22) and Nigeria (11) hosting about half of these (ENF, 2022). There might be many more unregistered recycling facilities across the continent. Furthermore, many recycling activities are semi-informal, characterised by suboptimal equipment and technologies. This wide gap in recycling facilities has fostered the creation of several small-scale enterprises aimed at tackling the challenge by using plastic waste as an economic resource (Oyinlola et al., 2022). These small and medium enterprises (SMEs) have attracted a growing support in the waste management value chain as it creates opportunities for collaboration to support a social, economic and environmental challenge. These organisations have increasingly received support from key actors such as local and foreign governments, investors, donor organisations and multinational companies, among others. They are also partnering with other actors in the value chains – e.g., the collection and disposal sector and recyclers – to facilitate sustainable waste management of plastics (Lane, 2018).

Decentralised models are gaining traction and are being adopted across Africa, especially by small-scale enterprises embracing the use of technology in waste management. The Ugandan approach to waste management was changed upon the acknowledgement that the country lacked the ability to operate a centralised operation to cater for the environmental and community requirements. A decentralised policy was therefore enacted in 1997 and further developed by the Local Government Act (Okot-Okumu and Nyenje, 2011). In addition, some countries in East Africa have transitioned from predominantly centralised models to a combination of public and private approaches with the inclusion of various stakeholders, principally service providers covering the diverse urban locations (Okot-Okumu, 2012).

A LMDCE gives waste plastic an economic value, which incentivises people to collect and use it locally, reducing waste accumulation (Babaremu et al., 2022; Oyinlola and Whitehead, 2020). It ensures the collection, disposal, remanufacture and reuse of plastics are done within the community (Joshi et al., 2019). It further significantly reduces the need for physical and technical infrastructure to implement an industrial circular economy of plastic by involving local community participation. For example, most economically disadvantaged countries have an informal local recycling ecosystem via an organisation of waste pickers (Fergutz et al., 2011; Medina, 2008; Parker, 2018). Waste pickers navigate through rural and urban cities to collect recyclable materials such as metals, plastics, glass and paper from various households and drop-off points while paying a small fee as incentive, before cleaning and sorting to further resell for a profit (Joshi et al., 2019). This process allows dense and heavily populated communities the opportunity to benefit from a decentralised circular economy, which includes plastic recycling. The circulation of plastics facilitates the replacement of the produce–consume–discard model as it develops a manufacturing supply chain by promoting using, recycling or re-entering on an industrial scale (Kaur et al., 2018; Kirchherr et al., 2017).

2.1 Overview of Digital Technologies for Waste Recycling in Africa

As highlighted previously, several small-scale enterprises have sprung up across the continent who are using digital technology such as mobile applications, geographic information systems (GIS) and artificial intelligence (AI). Oyinlola et al. (2022)

presented a summary of some of these organisations highlighting what digital technologies they use. It was observed that the common model adopted by these emerging enterprises includes at least one of the following components:

- Subscription: Majority of these start-ups have a database which is populated when customers within local communities sign up. Analysing the data gathered from these companies shows that these community-level subscriptions are usually regular, but in some cases, the interactions can be one-off. The level of interactions between the companies and subscribers is largely a function of the design of the communication campaigns, the incentives for participation and engagement and outreach activities designed by the companies. Typical customers include household, businesses, waste pickers and waste collectors. This database keeps information that allows the organisation to deliver its services, such as providing incentives, scheduling waste pickups and charging customers. Some of the start-ups (e.g., GIVO) are using their subscriptions to provide a mailing list and initiate a two-way communication with their end-users to have a better understanding of how to improve performance and services or run consumer behaviour surveys and collect market research data.
- Collection: This involves collecting recyclable waste from waste producers (including downstream users and corporate partners). Collection is done by various means including scavengers, bicycles, tricycles and mini electric vans. The quantity and type of recyclable waste collected from each collector are measured and recorded to enable incentives to be properly calculated. In recent years, new technology-based companies have adopted global positioning system (GPS) technology to track collection journeys with the aim of operational optimisation and gaining quantitative data on plastic feedstock across different communities.
- Processing: The recyclable plastics collected are sorted based on the type of plastic [polyethylene terephthalate (PET), high-density polyethylene (HDPE), polyvinyl chloride (PVC), etc.]. Depending on the feedstock condition, some of the recycling operations take the plastics through a washing and cleaning process. However, most of the decentralised plastic waste management systems try to avoid the cleaning and washing process by engaging with their communities and informing them about the feedstock quality suitable for their operations. The important stage for the mechanical recycling of the plastic is the shredding which converts the plastic waste into higher value products. The survey of the companies studied here shows that shredding can be problematic in the context of decentralised plastic waste recycling in sub-Saharan Africa. The main issue is the high cost associated with the operation and maintenance of the shredders and the scarcity of technical capacity to conduct the regular checks and services. Lack of appropriate training for staff prior to processing operations often results in overuse and overloading of mechanical equipment which often increase the maintenance need and operation costs for these businesses. Following the shredding, the recyclates will be sold to uptakers who recycle plastics.

3 Discussion

This decentralised plastic recycling approach adopted by the young tech-based entrepreneurs in sub-Saharan Africa is leading to wider societal impacts, as they mostly employ a community franchise model owned and operated by marginalised communities within the society, including women and youth in leadership positions. This approach to decentralised plastic recycling solution can be enhanced by the principle of a locally operated decentralised economy. The principle ensures suitable technology is applied to use available raw materials to manufacture goods based on accessible resources which would in turn ease the issue of unmanaged and mismanaged plastics (Browning et al., 2021). In the case of managing plastic waste, this translates to adding value to the processed recyclables by converting them into useful products. Decentralised solutions also aid efficiency and improve living of inhabitants due to localisation of processes in both rural and urban communities. These benefits can be observed in other sectors such as the energy sector with the use of decentralised hybrid photovoltaic (PV) solar-diesel power systems (Adaramola et al., 2014), the decentralised renewable energy systems (Oyedepo et al., 2018) and the health sector with decentralised health systems (Abimbola et al., 2015). Decentralised solutions have led to improved service delivery, democracy and participation and reduction in the central government's expenditure (Khan Mohmand and Loureiro, 2017).

Even though most of these organisations are embracing technology in plastic waste management, there is still significant scope to utilise technologies that deliver innovation across all aspects of the plastic value chain. For example, Chidepatil et al. (2020) suggested that AI drawing on multiple sensors and backed by the traceability of blockchain could remove barriers to a circular plastic economy. They argue that the use of AI can segregate plastic waste, therefore ensuring efficient and intelligent segregation, which is currently an inefficient process. They further argued that blockchain technology would be a useful platform for a trusted exchange across the value chain as it allows the information to be easily exchanged and validated along the value chain, providing different partners with relevant information on plastic waste and how best to reduce or recycle it. Singh (2019) illustrated how municipal waste management can make use of GIS and the layers available from remote sensing, while Mdukaza et al. (2018) highlighted the use of Internet of things (IoT) in plastic waste management. Other scholars including Hoosain et al. (2020), Kristoffersen et al. (2020) and Schot and Kanger (2018) have provided insights on how technology could be used to enhance waste management. Oyinlola et al. (2022) and Kolade et al. (2022b) have identified ten different technologies that could accelerate the transition to a circular plastic economy. Table 13.1 presents some of these technologies and highlights how they could enhance the productivity of these organisations.

TABLE 13.1 Digital technologies for the circular plastic economy

Digital Technologies	Functionality	Benefits
AI	Identification of plastic waste	Optimise circularity across the entire circular plastic economy (CPE) ecosystem
GIS	Geolocation of waste and connecting collectors to aggregators	Streamline operations in the CPE as well as efficiently connect CPE stakeholders
Blockchain	Capture of the lifecycle/journey of a plastic production	Foster transparency and facilitate data exchange across the CPE
IoT	- Waste identification and reporting to a central database via smart bins - Automated data collection from sensors - Conversion of recycled materials to finished and semi-finished products	Support embedding sensors for information exchange across the CPE
Robotics	Assisted waste sorting	Support automation across the CPE
3D printing	Repurposing plastic waste for filament production	Support decentralised recycling and reuse in the CPE
Function as a service (FaaS)	- Scalable solutions deployment - Pay-as-you-use model for infrastructure need - Digital innovations (DIs) focus more on their innovation rather than support systems	Eliminate the cost of infrastructural setup and deployment
Augmented reality/virtual reality (AR/VR)	Building digital solutions	Aid building digital solutions for awareness, sensitisation and training on best practices
5G	Real-time communication from collection centres and IoT sensors	Support real-time communication using IoT sensors
Mobile apps	- Data collection from source, information dissemination - Aggregation of data - Reward system implementation for collectors - Scheduling of waste pickup	Serve as an essential interface for all CPE stakeholders to interact for circularity

This shows that there is an opportunity to use a wide range of technologies to support the operations of decentralised plastic waste management enterprises. Some of these include the following:

- Mobile applications: for example, apps for collectors which can be used in conjunction with the hardware devices to manage the collection process, while customer apps would enable the customer to conveniently request pick up of their recyclables as well as view their historical deposits, impact of their deposit activities and incentives due. In summary, mobile apps are an essential interface for communicating across the value chain.
- IoT: Devices utilised for the collection and processing of recyclables can be integrated with IoT technology. This helps to digitise the entire process, facilitate mobile payments for recyclables collected and collate data that can be used to highlight waste consumption patterns. This also contributes to data for the entire process and is vital if blockchain is to be used. Examples of hardware devices that will benefit from IoT include scales, shredders, vehicles, etc.
- AI: This can be used for computer vision which can be utilised to identify the recyclables collected by colour, weight and brand. The computer vision technology tracks the recyclables from the collection point to the final recycled finished product, ensuring traceability and transparency of the waste management process.
- Cloud server: Information gathered from the IoT and AI enabled hardware devices can then be transmitted to the cloud server and processed in real time, for seamless record keeping and database management.
- Mobile payments: Facilitation of immediate, seamless mobile payments, as incentives for recyclables collected, to target users via mobile phones.

4 Conclusion

The future of technology within the plastic waste management system looks promising as it would foster greater interconnectedness between all stakeholders across the plastic value chain and traceability of materials collected and processed. This would in turn accelerate the transition to a more sustainable future. The application of technology such as the use of blockchain in data collection would help to make real change and optimise and inform the waste collection process. By collecting data at every stage of the plastic management cycle, the traceability of waste plastics from the source to the final recycled product is clearer. This provides valuable information about the waste streams which when applied to the circular economy principles can inform the processing, sale and repurposing of plastic goods to create a valuable economy for recyclables in Africa. Adopting digital technologies would reduce the reliance on people which could be seen as problematic in terms of jobs; however, there is an opportunity to upskill personnel in this space especially because a major challenge is the local procurement and maintenance of essential equipment such as eco-friendly and low-cost grinders and shredders with low-carbon footprints. These devices are usually sourced abroad and become obsolete once they develop a fault as there is limited local knowledge to fix them, and it is too expensive to send back to the originating country for repairs. Therefore, upskilling the current workforce to locally fabricate and maintain equipment will provide a low-cost solution that can be easily maintained over an extended period of time. It also empowers local artisans and fabricators, thereby leaving a positive footprint on the local economies of host communities.

In conclusion, decentralised plastic waste management solutions offer significant social, economic and environmental benefits to key stakeholders within the value chain. To accelerate the transition to a circular plastic economy, technological solutions need to be more modular, self-sustaining and efficient. These are necessary for effective management of the growing menace of plastic waste fuelled by linear plastic consumption culture, inadequate waste management infrastructure and unsustainable packaging activities of manufacturing and servicing organisations. Decentralised plastic waste management therefore needs to be underpinned by local sensitisation and public awareness about digital technologies for localised recycling across communities.

Acknowledgement

This work was supported by the Sustainable Manufacturing and Environmental Pollution (SMEP) programme funded with UK aid from the UK government Foreign, Commonwealth and Development Office (FCDO) (IATI reference number GB-GOV-1-30012).

References

Abimbola, S., Olanipekun, T., Igbokwe, U., Negin, J., Jan, S., Martiniuk, A., Ihebuzor, N., Aina, M., 2015. How decentralisation influences the retention of primary health care workers in rural Nigeria. *Glob. Health Action* 8, 26616.

Achankeng, E., 2003. Globalization, urbanization and municipal solid waste management in Africa, in: Proceedings of the African Studies Association of Australasia and the Pacific 26th Annual Conference. pp. 1–22.

Adaramola, M.S., Paul, S.S., Oyewola, O.M., 2014. Assessment of decentralized hybrid PV solar-diesel power system for applications in Northern part of Nigeria. *Energy Sustain. Dev.* 19, 72–82.

Adedibu, A.A., Okekunle, A.A., 1989. Issues in the environmental sanitation of Lagos mainland. *Nigeria. Environmentalist* 9, 91–100.

Anwar, S., Elagroudy, S., Abdel Razik, M., Gaber, A., Bong, C.P.C., Ho, W.S., 2018. Optimization of solid waste management in rural villages of developing countries. *Clean Technol. Environ. Policy* 20, 489–502.

Ayeleru, O.O., Dlova, S., Akinribide, O.J., Ntuli, F., Kupolati, W.K., Marina, P.F., Blencowe, A., Olubambi, P.A., 2020. Challenges of plastic waste generation and management in sub-Saharan Africa: A review. *Waste Manag.* 110, 24–42.

Babaremu, K.O., Okoya, S.A., Hughes, E., Tijani, B., Teidi, D., Akpan, A., Igwe, J., Karera, S., Oyinlola, M., Akinlabi, E.T., 2022. Sustainable plastic waste management in a circular economy. *Heliyon* 8, e09984. https://doi.org/10.1016/j.heliyon.2022. e09984

Baran, B., 2020. Plastic waste as a challenge for sustainable development and circularity in the European Union. *Ekon. i Prawo.Economics Law* 19, 7–20.

Barnes, D.K.A., Galgani, F., Thompson, R.C., Barlaz, M., 2009. Accumulation and fragmentation of plastic debris in global environments. *Philos. Trans. R. Soc. B Biol. Sci.* 364, 1985–1998.

Barrie, J., Anantharaman, M., Oyinlola, M., Schröder, P., 2022. The circularity divide: What is it? And how do we avoid it? *Resour. Conserv. Recycl.* 180, 106208. https://doi.org/10.1016/J.RESCONREC.2022.106208

Bläsing, M., Amelung, W., 2018. Plastics in soil: Analytical methods and possible sources. *Sci. Total Environ.* 612, 422–435.

Break Free From Plastic, 2019. BRANDED in Search of the World's Top Corporate Plastic Polluters, Acesso em, 15(11). https://www.breakfreefromplastic.org/wp-cont ent/uploads/2020/07/branded-2019.pdf

Browning, S., Beymer-Farris, B., Seay, J.R., 2021. Addressing the challenges associated with plastic waste disposal and management in developing countries. *Curr. Opin. Chem. Eng.* 32, 100682.

Chidepatil, A., Bindra, P., Kulkarni, D., Qazi, M., Kshirsagar, M., Sankaran, K., 2020. From trash to cash: How blockchain and multi-sensor-driven artificial intelligence can transform circular economy of plastic waste? *Adm. Sci.* 10, 23.

Drzyzga, O., Prieto, A., 2019. Plastic waste management, a matter for the 'community.' *Microb. Biotechnol.* 12, 66.

ENF, 2022. Plastic recycling plants in Africa – ENF recycling directory [WWW document]. www.enfrecycling.com/directory/plastic-plant/Africa accessed 10.18.22).

Fergutz, O., Dias, S., Mitlin, D., 2011. Developing urban waste management in Brazil with waste picker organizations. *Environ. Urban.* 23, 597–608.

Galloway, T.S., Lewis, C.N., 2016. Marine microplastics spell big problems for future generations. *Proc. Natl. Acad. Sci.* 113, 2331–2333.

Garmulewicz, A., Holweg, M., Veldhuis, H., Yang, A., 2016. Redistributing material supply chains for 3D printing. Proj. Report. Available online www.ifm.eng.cam. ac.uk/uploads/Research/TEG/Redistributing_material_supply_Chain.pdf (Accessed 16 July 2019).

Geyer, R., Jambeck, J.R., Law, K.L., 2017. Production, use, and fate of all plastics ever made. *Sci. Adv.* 3, e1700782. https://doi.org/10.1126/sciadv.1700782

Henry, R.K., Yongsheng, Z., Jun, D., 2006. Municipal solid waste management challenges in developing countries – Kenyan case study. *Waste Manag.* 26, 92–100.

Hoornweg, D., Bhada-Tata, P., Kennedy, C., 2013. Environment: Waste production must peak this century. *Nature* 502, 615–617.

Hoosain, M.S., Paul, B.S. Ramakrishna, S., 2020. The impact of 4IR digital technologies and circular thinking on the United Nations sustainable development goals. *Sustainability*, 12(23), 10143.

Hopewell, J., Dvorak, R., Kosior, E., 2009. Plastics recycling: Challenges and opportunities. *Philos. Trans. R. Soc. B Biol. Sci.* 364, 2115–2126.

Hornik, J., Cherian, J., Madansky, M., Narayana, C., 1995. Determinants of recycling behavior: A synthesis of research results. *J. Socio. Econ.* 24, 105–127.

IISD, 2019. Companies launch African plastics recycling alliance [WWW document]. SDG Knowl. HUB. https://sdg.iisd.org/news/companies-launch-african-plastics-recycling-alliance/ (accessed 10.18.22).

Jambeck, J.R., Geyer, R., Wilcox, C., Siegler, T.R., Perryman, M., Andrady, A., Narayan, R., Law, K.L., 2015. Plastic waste inputs from land into the ocean. *Science (80-.). 347*, 768–771.

Joshi, C., Seay, J., Banadda, N., 2019. A perspective on a locally managed decentralized circular economy for waste plastic in developing countries. *Environ. Prog. Sustain. Energy* 38, 3–11.

Joshi, C.A., Seay, J.R., 2020. Total generation and combustion emissions of plastic derived fuels: A trash to tank approach. *Environ. Prog. Sustain. Energy* 39. doi:10.1002/ep.13151

Kaseva, M.E., Mbuligwe, S.E., 2005. Appraisal of solid waste collection following private sector involvement in Dar es Salaam city, Tanzania. *Habitat Int.* 29, 353–366.

Kaur, G., Uisan, K., Ong, K.L., Ki Lin, C.S., 2018. Recent trends in green and sustainable chemistry & waste valorisation: Rethinking plastics in a circular economy. *Curr. Opin. Green Sustain. Chem.* 9, 30–39. https://doi.org/10.1016/j.cogsc.2017.11.003

Khan Mohmand, S., Loureiro, M., 2017. Interrogating Decentralisation in Africa. IDS Bulletin. 48, 2[11]. https://opendocs.ids.ac.uk/opendocs/handle/20.500.12413/12876. DOI: 10.19088/1968-2017.110

Kirchherr, J., Reike, D., Hekkert, M., 2017. Conceptualizing the circular economy: An analysis of 114 definitions. *Resour. Conserv. Recycl.* 127, 221–232. https://doi.org/10.1016/j.resconrec.2017.09.005

Kolade, O., Adegbile, A., Sarpong, D., 2022a. Can university-industry-government collaborations drive a 3D printing revolution in Africa? A triple helix model of technological leapfrogging in additive manufacturing. *Technol. Soc.* 69, 101960. https://doi.org/10.1016/j.techsoc.2022.101960

Kolade, O., Odumuyiwa, V., Abolfathi, S., Schröder, P., Wakunuma, K., Akanmu, I., Whitehead, T., Tijani, B., Oyinlola, M., 2022b. Technology acceptance and readiness of stakeholders for transitioning to a circular plastic economy in Africa. *Technol. Forecast. Soc. Change* 183, 121954. https://doi.org/10.1016/J.TECHFORE.2022.121954

Kreiger, M.A., Mulder, M.L., Glover, A.G., Pearce, J.M., 2014. Life cycle analysis of distributed recycling of post-consumer high density polyethylene for 3-D printing filament. *J. Clean. Prod.* 70, 90–96. https://doi.org/10.1016/J.JCLEPRO.2014.02.009

Kristoffersen, E., Blomsma, F., Mikalef, P., Li, J., 2020. The smart circular economy: A digital-enabled circular strategies framework for manufacturing companies. *J. Bus. Res.* 120, 241–261.

Lane, W., 2018. *Oceans of Plastics: Developing Effective African Policy Responses*. South African Institute of International Affairs.

Mdukaza, S., Isong, B., Dladlu, N., Abu-Mahfouz, A. M. 2018. Analysis of IoT-Enabled Solutions in Smart Waste Management, IECON 2018 - 44th Annual Conference of the IEEE Industrial Electronics Society, Washington, DC, USA, pp. 4639–4644, doi: 10.1109/IECON.2018.8591236

Medina, M., 2008. *The Informal Recycling Sector in Developing Countries: Organizing Waste Pickers to Enhance Their Impact*. World Bank.

Narancic, T., O'Connor, K.E., 2019. Plastic waste as a global challenge: Are biodegradable plastics the answer to the plastic waste problem? *Microbiology* 165, 129–137.

OECD, 2022. Global plastics outlook [WWW document]. OECD iLibrary. www.oecd-ilibrary.org/environment/data/global-plastic-outlook_c0821f81-en (accessed 11.19.22).

Okot-Okumu, J., 2012. Solid waste management in African cities–East Africa. In Fernando, L., Rebellon, M. (Eds.), *Waste Management – An Integrated Vision* (pp. 3–20). IntechOpen.

Okot-Okumu, J., Nyenje, R., 2011. Municipal solid waste management under decentralisation in Uganda. *Habitat Int.* 35, 537–543.

Oyedepo, S.O., Babalola, P.O., Nwanya, S., Kilanko, O.O., Leramo, R.O., Aworinde, A.K., Adekeye, T., Oyebanji, J.A., Abidakun, O.A., Agberegha, O.L., 2018. Towards a sustainable electricity supply in Nigeria: The role of decentralized renewable energy system. *Eur. J. Sustain. Dev. Res.* 2, 40.

Oyinlola, M., Kolade, S., Odumuyiwa, V., Schröder, P., Whitehead, T., Wakunuma, K., Lendelvo, S., Rawn, B., Sharifi, S., Akanmu, I., Brighty, G., Mtonga, R., Tijani, B., Abolfathi, S., 2022. A socio-technical perspective on transitioning to a circular plastic economy in Africa. . *SSRN Electron. J.* https://doi.org/10.2139/ssrn.4332904

Oyinlola, M., Okoya, S.A., Whitehead, T., Evans, M., Lowe, A.S., 2023b. The potential of converting plastic waste to 3D printed products in Sub-Saharan Africa. *Resour. Conserv. Recycl. Adv.* 17, 200129. https://doi.org/10.1016/j.rcradv.2023.200129

Oyinlola, M., Schröder, P., Whitehead, T., Kolade, S., Wakunuma, K., Sharifi, S., Rawn, B., Odumuyiwa, V., Lendelvo, S., Brighty, G., Tijani, B., Jaiyeola, T., Lindunda, L., Mtonga, R., Abolfathi, S., 2022. Digital innovations for transitioning to circular plastic value chains in Africa. *Africa J. Manag.* 8, 83–108. https://doi.org/10.1080/23322 373.2021.1999750

Oyinlola, M., Whitehead, T., 2020. Recycling of plastics for low cost construction. In Hashmi, S., Choudhury, I.A.B.T.-E. of R. and S.M. (Eds.), *Reference Module in Materials Science and Materials Engineering* (pp. 555–560). Elsevier. https://doi.org/10.1016/B978-0-12-803581-8.11523-1

Oyinlola, M., Whitehead, T., Abuzeinab, A., Adefila, A., Akinola, Y., Anafi, F., Farukh, F., Jegede, O., Kandan, K., Kim, B., Mosugu, E., 2018. Bottle house: A case study of transdisciplinary research for tackling global challenges. *Habitat Int.* 79, 18–29. https://doi.org/10.1016/j.habitatint.2018.07.007

Parker, L., 2018. Planet or plastic: Fast facts about plastic pollution. *Natl. Geogr. Mag*, 20. https://www.nationalgeographic.com/science/article/plastics-facts-infographics-ocean-pollution

Plastics Europe, 2018. Plastics – the facts 2018. An analysis of European plastics production, demand and waste data. Plastics Europe. Available at: https://plasticseurope.org/ (Accessed: 19 March 2021)

Ragaert, K., Delva, L., Van Geem, K., 2017. Mechanical and chemical recycling of solid plastic waste. *Waste Manag.* 69, 24–58.

Ritchie, H., Roser, M., 2018. *Plastic Pollution*. Our World Data.

Rochman, C.M., Browne, M.A., Halpern, B.S., Hentschel, B.T., Hoh, E., Karapanagioti, H.K., Rios-Mendoza, L.M., Takada, H., Teh, S., Thompson, R.C., 2013. Classify plastic waste as hazardous. *Nature* 494, 169–171.

Rockström, J., Steffen, W., Noone, K., Persson, Å., Chapin III, F.S., Lambin, E., Lenton, T.M., Scheffer, M., Folke, C., Schellnhuber, H.J., 2009. Planetary boundaries: Exploring the safe operating space for humanity. *Ecol. Soc.* 14, 32.

Santander, P., Cruz Sanchez, F.A., Boudaoud, H., Camargo, M., 2020. Closed loop supply chain network for local and distributed plastic recycling for 3D printing: A MILP-based optimization approach. *Resour. Conserv. Recycl.* 154, 104531. https://doi.org/10.1016/J.RESCONREC.2019.104531

Schot, J., Kanger, L., 2018. Deep transitions: Emergence, acceleration, stabilization and directionality. *Res. Policy.* 47(6), 1045–1059.

Schroeder, P., Oyinlola, M., Barrie, J., Bonmwa, F., Abolfathi, S., 2023. Making policy work for Africa's circular plastics economy. *Resour. Conserv. Recycl.* 190, 106868. https://doi.org/10.1016/j.resconrec.2023.106868

Schumacher, E.F., 2011. *Small Is Beautiful: A Study of Economics as If People Mattered.* Random House.

Shen, M., Huang, W., Chen, M., Song, B., Zeng, G., Zhang, Y., 2020. (Micro)plastic crisis: Un-ignorable contribution to global greenhouse gas emissions and climate change. *J. Clean. Prod.* 254, 120138. https://doi.org/10.1016/j.jclepro.2020.120138

Singh, A., 2019. Remote sensing and GIS applications for municipal waste management. *J. Environ. Manage.* 243, 22–29. https://doi.org/10.1016/j.jenvman.2019.05.017

Singh, P., Sharma, V.P., 2016. Integrated plastic waste management: Environmental and improved health approaches. *Procedia Environ. Sci.* 35, 692–700.

Thøgersen, J., 1996. Recycling and morality: A critical review of the literature. *Environ. Behav.* 28, 536–558.

Thompson, R.C., Moore, C.J., Vom Saal, F.S., Swan, S.H., 2009. Plastics, the environment and human health: Current consensus and future trends. *Philos. Trans. R. Soc. B Biol. Sci.* 364, 2153–2166.

UNEP, 2018a. *Mapping of Global Plastics Value Chain and Plastics Losses to the Environment: With a Particular Focus on Marine Environment.* United Nations Environment Programme, Nairobi.

UNEP, 2018b. *Africa Waste Management Outlook.* Nairobi.

Vanapalli, K.R., Sharma, H.B., Ranjan, V.P., Samal, B., Bhattacharya, J., Dubey, B.K., Goel, S., 2021. Challenges and strategies for effective plastic waste management during and post COVID-19 pandemic. *Sci. Total Environ.* 750, 141514.

Wan, M., Wan, L., 2020. Exploring the pathways to participation in household waste sorting in different national contexts: A fuzzy-set QCA approach. *IEEE Access* 8, 179373–179388.

Waters, C.N., Zalasiewicz, J., Summerhayes, C., Barnosky, A.D., Poirier, C., Gałuszka, A., Cearreta, A., Edgeworth, M., Ellis, E.C., Ellis, M., 2016. The Anthropocene is functionally and stratigraphically distinct from the Holocene. *Science (80-.).* 351, aad2622.

Wilcox, C., Van Sebille, E., Hardesty, B.D., 2015. Threat of plastic pollution to seabirds is global, pervasive, and increasing. *Proc. Natl. Acad. Sci.* 112, 11899–11904.

Williams, M., Gower, R., Green, J., Whitebread, E., Lenkiewicz, Z., Schröder, P., 2019. *No Time to Waste: Tackling the Plastic Pollution Crisis before It's too Late.* Tearfund.

Wilson, D.C., Rodic, L., Modak, P., Soos, R., Carpintero, A., Velis, K., Iyer, M., Simonett, O., 2015. *Global Waste Management Outlook.* UNEP.

Wilson, D.C., Velis, C.A., Rodic, L., 2013. Integrated sustainable waste management in developing countries, in: Proceedings of the Institution of Civil Engineers-Waste and Resource Management. ICE Publishing, pp. 52–68.

Wright, S.L., Kelly, F.J., 2017. Plastic and human health: A micro issue? *Environ. Sci. Technol.* 51, 6634–6647.

Yhdego, M., 1995. Urban solid waste management in Tanzania Issues, concepts and challenges. *Resour. Conserv. Recycl.* 14, 1–10.

Zhong, S., Pearce, J.M., 2018. Tightening the loop on the circular economy: Coupled distributed recycling and manufacturing with Recyclebot and RepRap 3-D printing. *Resour. Conserv. Recycl.* 128, 48–58. https://doi.org/10.1016/J.RESCONREC.2017.09.023

14

ASSESSING PLASTIC CIRCULAR ECONOMY POLICIES AND THE USE OF DIGITAL TECHNOLOGY IN AFRICA

Olubunmi Ajala

1 Introduction

The world is confronted with complex and interconnected environmental problems (Kinzig et al., 2013). These problems range from ozone depletion, climate disruption, species declines and extinctions, emerging diseases, antibiotic resistance, persistent organic pollutants, amongst others. Amongst these major challenges is plastic pollution which has been attracting increasing attention lately (Syberg et al., 2021). While extensive studies exist on how to reduce plastic waste (Austin et al., 1993; Ayeleru et al., 2020; Chow et al., 2017; Luo et al., 2022; Rochman et al., 2013), there is a shortage of materials evaluating the effectiveness of numerous initiatives that have been proposed to tackle plastic waste problems. In addition to that, there is a shortage of literature on how these initiatives are relevant to Africa's unique situations and in transforming Africa's plastic "throwaway" economic model. In this chapter, we review some of these regulations within an African circular economic framework and assess how digital innovations can help in reducing plastic waste on the continent.

There has been an increasing interest in the growing prevalence of (micro) plastics in the environment, particularly regarding its effects on marine ecosystems. This increased attention over ocean life has led to the implementation of some strict guidelines and policies, but the efficacy of these regulations remains widely undetermined (da Costa et al., 2020). Similarly, because there is no convention with the sole aim of solving the plastic waste problem (it is always an integral part of other pollution-related regulations), there are no unified and integrated mechanisms regulating and controlling the spread of plastic materials.

To further increase interest in plastic waste problem is the COVID-19 pandemic. Just as most health shocks tend to produce some negative economic outcomes (Alam

DOI: 10.4324/9781003278443-17

& Mahal, 2014), COVID-19 pandemic has been asserted to have also exacerbated the plastic waste challenge as the use of single-use plastic products seemed to have tripled during the pandemic, thereby providing further justification for looking at this topic.

In this development, Africa is not insulated from these challenges, rather Africa equally has a significant role to play in taming environmental degradation. Africa holds a topical position in discussing marine ecosystems because two of the ten top-ranked rivers (Niger and Nile) that transport 88–95% of the global plastic waste load into the sea flow through Africa (Schmidt et al., 2017). It is also projected that the volume of wastes in Sub-Saharan Africa will almost double by 2050, with much of this being plastic (Kaza et al., 2018). Ayeleru et al. (2020) estimates that 17 Mt of plastic waste is annually generated in Africa, hereby calling for urgent attention. It has been said that inaction may result in the number of plastics in the ocean exceeding the number of fish by 2050. According to Ellen MacArthur Foundation, by 2040, the circular economy has the potential to reduce annual plastic inflow into oceans by 80%, which can cut greenhouse emissions by 25%, create about 700,000 net additional jobs and generate over USD 200 bn per year in savings.

From our observation, it is difficult finding specific circular economy legislations or legislations particularly focusing on plastic waste circular economy. It becomes more difficult finding policies that are circular and are aimed at plastic economy. Our observation aligns with Desmond and Asamba (2019) in which they consider one country (Rwanda) out of eight selected countries with circular economy initiatives directly focusing on plastic waste. This has also translated into dearth of systematic studies of circular economy policies in Africa, and there is a massive research gap in identifying the extent and impact of sustainability policies regarding plastic wastes in Africa. In this chapter, we annotate on some international initiatives that were designed to influence plastic waste management on the continent before delving into a case study of policy state in Nigeria.

Review of studies by Cagno et al., (2021) shows that despite the availability of literature on the broad topic of circular economy there is limited studies on digital technologies in the circular economy transition (Cagno et al., 2021). Similarly, researchwise, focus on the plastic circular economy is in its infancy for the most part of Africa (Oyinlola et al., 2022).

In this chapter, we use a mixed method of analysis. We utilise machine learning to undertake text analysis of policy descriptions across Africa. We also undertake a descriptive data analysis of the DITCh Plastic Survey which is akin to Facebook-Yale climate change opinion survey and the WWF-SA study survey (South Africa) to assess people's perception of plastic waste and plastic waste policies in Africa. Our survey includes individual's levels of awareness of plastic waste policies and perceptions of plastic waste policy effectiveness on the continent. Insight from the data is combined with national case studies of two policies in Nigeria.

We find that the conception and drafting of legislation are shallow such that there is incomplete approach to tackling plastics waste problems, there is lack of awareness of the initiatives, exclusion of the informal sector and the existence of enforcement problems such as lack of information on usage and traceability. In our attempt to investigate how digital tools and innovations could enhance policy implementation for reducing plastic waste, we also reviewed the DITCh Plastic digital innovation aggregator, which presents some of the technological start-ups' ideas in Africa. We then propose practical digital technological tools that can enhance plastic circular economy policies in Africa. We find a hollow implementation of circular plastic economy initiatives in Africa, with little evidence of their success (Kweku & Johanna, 2020). We reiterate the relevance of our conclusion for policy (from the text analysis that we conducted), i.e., that African initiatives need to move beyond prohibition into a circular framework (re-use, re-cycle, re-make, etc.).

We support the World Economic Forum (2021b) assertion that governments should set up political, legislative and economic frameworks that can incentivise profitable circular economies by facilitating a digital backbone for Africa and support the claim that digital innovations can generate economies of scale (World Economic Forum, 2021a) for circular plastic economy stakeholders by connecting stakeholders from a wide range of backgrounds, sectors and countries across Africa (Oyinlola et al, 2022) and that digital innovations can be used to aid recycling by efficiently connecting consumers, waste collectors and recyclers; reduce plastic wastes by engaging consumers on ways to cut down resource usage; and aid redesign by optimising processes.

Finally, countries have adopted various approaches to stem environmental degradation such as the use of education, persuasion and policies, but judging from the low success level of current efforts, there is the need to review current initiatives particularly in Africa where resource-rich syndrome may impede performance-based ideas, profit-maximisation objectives may hinder sustainability consideration and prominent roles of the informal sector are often ignored in policies. However, before delving into discussing some of the plastic waste policies in Africa, there is the need to briefly discuss key economic concepts relevant to our study.

Section 1 of this chapter presents the general introduction. Section 2 introduces the concept of circular economy. Section 3 focuses on the plastic circular economy policies in Africa. Section 4 presents the descriptive statistics from our survey and looks at the role digital innovation can play in scaling the circular economy in Africa. Section 5 presents the summary and conclusion of the chapter.

2   Concept of Circular Economy

The concept of circular economy has been around for a number of decades (Lacy & Rutqvist, 2016). It is an approach that keeps resources in productive use for as long as possible. A good analogy often used to explain a circular economy is comparing a river with a lake. A linear economy is depicted to flow like a river where products are created through a series of value-adding activities and upon sales of the product, the property right and the liability for risks and wastes are transferred to the buyer. The owner thereafter decides what happens to old products (discarded, reused or recycled). On the other hand, a circular economy is depicted as a lake where reprocessing of products takes place. It operates as a system where the objective is to maximise a product's value at each point of its life.

The contrast between an "open economy" where input is unlimited and a "closed economy" where resources are bounded was raised in the famous essay of Boulding (2011). Boulding's essay is usually cited as the origin of the concept of "circular economy". The concept introduces required "closing loops" into an open economy by ensuring that products that are at the end of their service life are either reused or turned into other resources, thereby minimising waste (Stahel, 2016). One critical linkage between plastic pollution and a circular economy is the fact that the traditional economic production cycle of "make, use, dispose" is said to have resulted in one-third of plastic wastes not collected or managed globally (MacArthur et al., 2016).

Ellen MacArthur Foundation (EMF) also reiterates that a circular economy is restorative and regenerative by design. It distinguishes between technical and biological cycles and aims at keeping products, components and materials at their highest value always. It is suggested that the transition from linear economy to a circular economy is the biggest revolution in 250 years as it presents a radical rethink of the relationship between customers, markets and natural resources while at the same time presents biggest opportunities (Lacy & Rutqvist, 2016).

Factors driving the circular economy adoption are principally resource constraints, technological developments and socio-economic opportunities (Lacy & Rutqvist, 2016). A study of seven European nations found that transitioning to a circular economy will reduce greenhouse gas emissions of each country by about 70% and grow their workforce by about 4% (Stahel, 2016). Recycling is a well-known element of the circular economy, but there are other elements that are not well-publicised yet. Our focus in this chapter is on how digital technologies can play important roles in scaling Africa's circular economy. When we talk about digital technology, we refer to innovations around cloud, mobile, social, big data analytics, internet of things (IoT), amongst others.

3 Circular Economy Policies in Africa

This segment attends to three key areas. The first part briefly reviews international marine initiatives that may likely have implications on Africa circular plastic economy. The second part of the segment delves deeply into initiatives in Africa that are considered under green economy that may have wider implications on Africa circular plastic economy. The last part looks at the state of national policy in one of the countries that is recorded to have one of the highest initiatives on green economy in Africa (Nigeria).

3.1 Global Marine Policies and Africa Cross-Border Plastic Initiatives

There are a few international frameworks in operation seeking to attend to plastic waste problems such as the United Nations Convention on the Law of the Sea (UNCLOS) of 1982. The law attempts to regulate every aspect of the sea resources and the use of the ocean (Gagain, 2012). Article 210 of the convention encourages individual states to develop frameworks to prevent, reduce and control pollution of the marine environment. However, as well-intended as the "Constitution for the Oceans" is, it focuses on a wide range of areas, but it did not specifically contain any provisions regarding plastic pollution. It rather considered plastic as all other wastes potentially hazardous for the ocean life (da Costa et al., 2020). The implication of this is a slack domain to directly tackle plastic waste as a core challenge. Similarly, the position of countries such as the United States not to be signatory to the law has reduced the effectiveness of UNCLOS in tackling plastic problems (Bateman, 2007).

A similar intervention, supported by the United States, the "Marine Debris Program" (MDP) specifically designed to curb marine debris was jointly developed by the United Nations Environment Program (UNEP) and the US National Oceanic and Atmospheric Administration (NOAA). The programme aimed at fighting the increasing prevalence of marine litter, but it also has its limitations in application because its functionality depends on the willingness of participating nations. Because of its non-binding nature, we view this as a tame attempt at tackling the challenge of plastic pollution at the global level.

Similarly, a resolution on marine litter and micro plastics was passed during the United Nations Environment Assembly (UNEA) of UNEP in Nairobi (Kenya in 2017), urging all countries to make responsible use of plastic while endeavouring to reduce unnecessary plastic use. UNEA-4 acknowledged the problem of micro plastics, marine plastic litter and the problem of single-use plastic. Various resolutions from UNEA are considered good global initiatives at understanding plastic solutions with the aim of informing global policies. As an extension of this, the African Ministerial Conference on the Environment (AMCEN) in Durban (South Africa 2019) has also passed a declaration emphasising the need to address plastic pollution.

A cross-border initiative involving Rwanda, Nigeria and South Africa announced the African Circular Economy Alliance (ACEA) during COP23 in Bonn as an effort to create inter-governmental corporations. One of their aims was to encourage other African countries to consider implementing similar policies as Nigeria's EPR programme (Desmond & Asamba, 2019). The initiative was also to facilitate knowledge sharing of the empirical applications of circular economy to different sectors of the economy between circular economy professionals in Africa. While the inclusion of the two biggest economies in Africa (Nigeria and

South Africa) in this initiative is significant, the number of countries involved shows one of the challenges of global initiatives (low participation level).

Review of these global and regional efforts shows some shortcomings as a limited number of countries participate in most of these initiatives and many of them are non-binding agreements. We are also of the view that their implementations also lack a compliance mechanism making accountability also difficult.

3.2 National Policies in Africa

Policy intervention has been identified as a critical driver of sustainable circular plastic economy (Dijkstra et al., 2020). On the positive side, our review of circular economy policies in Africa shows that there is small evidence of initiatives around climate change and green economy in general. At least, 36 African countries have introduced one form of initiatives or the other (see Figure 14.1, left part). It shows the density of national green economy policies across Africa. Policies by count shows some good representation, as the top economies in Africa in terms of gross domestic products (Nigeria, South Africa and Egypt) lead in terms of policy count while there are only three countries in Africa (Equatorial Guinea, Gibraltar and South Sudan) with no recorded evidence of green economy initiatives. The flip side of it is that they are mostly towards the banning of plastic bags (Kweku & Johanna, 2020).

We reviewed 241 policies or interventions in Africa but only eight fell under National Circular Economy (NCE), i.e., only 3% clearly fell under NCE (Table 14.1).

To highlight these policies, Figure 14.1 (right part) shows distribution of policy across Africa that are classified as "Circular Economy Policy" (Chatham House,

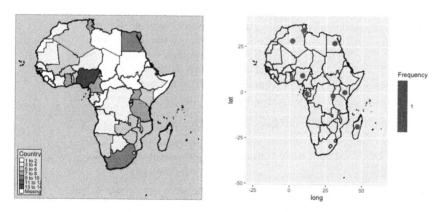

FIGURE 14.1 Density of national green economy policies and number of policy categorised as circular economy policy across Africa

Note: Data Sourced from Chatham House. Summarised data for generating the map 1 is in Appendix 1, map 2 is generated from Table 14.2

TABLE 14.1 Summary of green economy policies by category in Africa

Policy Category	Policy Count	Proportion
Extended Producer Responsibility EPR	26	11%
Fiscal Policy	14	6%
National Circular Economy Policy	8	3%
Product Policy	48	20%
Waste Management and Recycling	145	60%
Total	**241**	

Note: Data sourced from Chatham House database – https://circulareconomy.earth/

TABLE 14.2 Circular economy policies in Africa

Country	Policy Name	Policy Description	Start year
Algeria	National action plan for sustainable production and consumption methods (PNA-MCPD) 2016—2030	The plan has three overarching aims: 1) to integrate sustainable consumption and production patterns into national policies and plans; 2) to ensure energy transition through promotion of energy efficiency and 3) development of renewables and develop a zero-waste economy by 2030. Includes specific actions to accelerate the circular economy transition through greater recycling, improving waste management services, eco-design, and life cycle analysis.	2016
Egypt	National Action Plan for Sustainable Consumption and Production	This action plan aims to support Egypt's development efforts in achieving sustainable consumption and production practices in its key economic sectors, including energy, agriculture, water and waste. Regarding waste management, the action plan advocates for waste prevention, reduction, recycling, re-use and recovery. It also promotes a gradual transition to a green and circular economy as a conceptual framework for policy making. This strategy also highlights steps to be taken to promote a gradual adoption of governmental procurement towards environmentally friendly products and sustainable technologies.	2016
Gabon	Gabon Green Operational Plan	The plan sets the green strategy for Gabon, with the 'aim to increase the level of wealth produced while controlling the footprint ecological effects of human activities'. It specifically mentions the application of circular economy principles in the plan and the promotion of waste recycling channels.	2015

TABLE 14.2 (Continued)

Country	Policy Name	Policy Description	Start year
Kenya	The Green Economy Strategy and Implementation Plan (GESIP) (2016),	Green Economy Strategy and Implementation Plan 2016–2030 is geared towards enabling Kenya to attain a higher economic growth rate consistent with the Vision 2030, which firmly embeds the principles of sustainable development in the overall national growth strategy. This strategy builds on the achievements realised during the implementation of the first Medium Term Plan (MTP I 2008–2012) and on-going implementation of MTP II (2013–2017) for Vision 2030. The strategy aims to shift the attitudes of households and industry towards sustainable consumption and production and sustainability.	2016
Madagascar	Environmental Program for Sustainable Development	The programme has two strategic objectives broken down into six specific objectives; Strategic Objective 1: an effective environmental policy framework, an optimised environmental performance of development actors, and a reliable information system as a decision support device; and Strategic Objective 2: an inventory of natural capital and the benefits generated at a known national level, a network of green infrastructures managed effectively and increasing resilience to risks of disasters, and fair and equitable sharing of the benefits of nature strengthening socio-economic resilience, including objectives around waste valorisation and recycling.	2016
Nigeria	National Policy on the Environment (Revised 2016)	The goal of the National Policy on the Environment is to ensure environmental protection and the conservation of natural resources for sustainable developments. Its strategic objective is to coordinate environmental protection and natural resources conservation for sustainable development. Waste management is an important part of this policy, with a whole Objective focusing on "Waste and Environmental Pollution" and looks at solid waste, household, and industrial waste, wastewater, toxic and hazardous waste, radioactive waste.	2016

(Continued)

TABLE 14.2 (Continued)

Country	Policy Name	Policy Description	Start year
Rwanda	Rwanda National Environment and Climate Change Policy	The National Environment and Climate Change Policy provides strategic direction and responses to the emerging issues and critical challenges in environmental management and climate change adaptation and mitigation. The key issues and challenges identified include high population density, water, air and soil pollution, land degradation, fossil-fuel dependency, high-carbon transport systems, irrational exploitation of natural ecosystems, lack of low-carbon materials for housing and green infrastructure development, inadequate waste treatment for both solid and liquid waste, increase of electronic, hazardous chemicals and materials waste, among others. The policy includes seven policy objectives, of which Policy objective 1: Greening economic transformation includes a specific statement policy statement on promoting the circular economy.	Start year
Tunisia	National Strategy for the Green Economy 2016–2036	The purpose of the national strategy is to explore possibilities of development of current economic activity and new green activities in several areas, including organic farming and eco-tourism, sustainable transport and infrastructure, sustainable buildings and green industries, environmental services, energy efficiency and renewable energy, water conservation and water re-use and integrated waste treatment management. It includes focus areas 1 & 3 linked to the circular economy and waste management; 'cultivate efficiently in the use of natural resources, less polluting and the ocean with sustainable production' and 'waste disposal in an integrated framework in order to improve life by recovering recycled waste and reducing greenhouse gas emissions'.	2016

Note: Policy extracts and categorisation from (Chatham House, 2021)

TABLE 14.3 Countries and number of policy descriptions containing "plastic"

Countries	Policy Counts
Mauritius	5
Mali, Seychelles	4
Benin, Zimbabwe	3
Burkina Faso, Côte d'Ivoire, Gambia, Ghana, Namibia, São Tomé and Príncipe, Senegal, South Africa, Togo, Zambia	2
Algeria, Angola, Botswana, Burundi, Cameroon, Cape Verde, Democratic Republic of the Congo, Djibouti, Egypt, Eritrea, Ethiopia, Gabon, Guinea-Bissau, Kenya, Lesotho, Madagascar, Malawi, Mauritania, Morocco, Mozambique, Niger, Republic of the Congo, Rwanda, Somalia, Sudan, Tanzania, Tunisia, Uganda	1

Note: Data from Chatham House (2021)

TABLE 14.4 Distribution of policies that contained plastic by policy type

Policy Category	Policy Counts
Extended Producer Responsibility EPR	7
Fiscal Policy	9
Product Policy	42
Waste Management and Recycling	9
Total	**67**

Source: Data (Chatham House, 2021)

2021). Eight countries from all African countries recorded one initiative each considered to be circular in nature. The tabular presentation (Table 14.2) presents policies that are considered under circular economy with their descriptions and policy commencement date.

Delving deeper beyond this categorisation into individual policy description, we review the listed 241 policies to ascertain those who at least mentioned "plastic" in their descriptions. We found 67 policies across Africa. Table 14.3 shows countries and number of policies relating to plastics.

In the light of our topic, we reviewed the categorisation of the 67 policies, using the five categories of policies under Chatham House. None of the policies in Africa that at least contained plastic in the description falls under Circular Economy Policy (see Table 14.4).

To further review the 67 policies that contain "plastic" in their description, we undertake text analysis of their descriptions, using machine learning. We tokenise every word used in the description, using Gibss LDA method of topic modelling (Hornik & Grün, 2011). The top 10 topics are built on "bags", "finance", "law", "levy", "plastic", "approving", "decree", "management", "plan" and "presidential"

FIGURE 14.2   Word cloud of words used in plastic-related policies in Africa

Note: R package used for word cloud. Policy description extracted from Chatham House Database

(in descending order). These further validates our earlier realisation that these policies are product based and are rather prohibitive in nature. We further select the most frequently used words in the policies (words with frequency more than 200 in the tokenised document). Figure 14.2 presents the resulting word cloud from the text analysis. The conspicuous topic from the leading terms is "... national law bans packaging bag". These are closely followed by terms such as "prohibiting", "prohibition", "producer", "distribution", amongst others.

Some of our takeaway from these facts are, Africa is positively initiating policies toward green economy, however, only a few leans towards circular economy. Some initiatives are plastic-focussed but none of such is categorised as circular in nature. We state that the intersection between policies that are plastic focussed and policies that are national circular plastic economy is zero in Africa (as at time of this publication). The relevance of our conclusion for policy from the text analysis is that African initiatives need to move beyond prohibition into a circular framework (re–use, re–cycle, re–make, etc.).

3.3   State of National Plastic Policies: A Case of Nigeria

This segment presents our national-level case studies. We have briefly discussed international and continental efforts at tackling plastic waste problems above, but for a more succinct scenario analysis, we review the two main initiatives relating to plastics wastes management in Nigeria. Our choice of Nigeria is based on the high level of plastic wastes the country generates (Ajani & Kunlere, 2019), fact that it currently has more green economy initiatives more than any other country in Africa (see Appendix 1). In addition, the country's policies and regulations are close to what other African countries (Rwanda, Kenya, Gambia, Ghana, South Africa and Morocco) have formulated. Insights from this review are then combined with our Africa-wide survey to generalise the state of plastic waste policies.

3.3.1 Plastic Bags Prohibition Bill

The Nigerian National Assembly passed the Plastic Bags Prohibition Bill in May 2019. The bill prohibits the use, manufacture and importation of plastic bags for commercial and household packaging. The bill provides that any retailer that provides customers with plastic bags at a point of sale is guilty of an offence. Likewise, a person who manufactures plastic bags for reselling is equally guilty, just as someone who imports plastic bags either as a carryout bag or for resale. For individual offenders, it proposed a penalty of a fine of not more than N500,000 (Five Hundred Thousand Naira or $1,400 USD), or a prison term not exceeding three years, or both. For corporate offenders, companies shall be liable to a fine of not more than N5,000,000 (Five Million Naira or $14,000 USD).

Despite the passage of the bill by the House of Representatives, the Nigerian Senate is yet to pass the bill into law, and it has not been assented into law by the President, but we still find it useful to review the initiative as part of transforming Nigeria to a circular economy.

The policy is not situated within a goal- or mission-oriented policy framework as it looks more like a statement of prohibitions and corresponding penalties unlike in the UK where the plastic waste policy of taxing users is linked to a 25-year environmental plan. When a policy is not mission-oriented, it does not only stand the chance of fading away in the short run, but it also makes appraising its effectiveness difficult.

The outright ban of plastic bags itself without inculcating other market-based instruments (e.g., plastic taxes, subsidy and incentives) provides a good basis for policy failure. This one-way approach which is similar to plastic ban in Kenya, Rwanda, Gambia and Morocco is shallow, as it does not address other viable plastic management options. A system approach would need to include aspects of recyclability and reusability if the continent is to veer towards circular plastic economy.

3.3.2 Nigeria Extended Producers Responsibility Programme

Germany introduced a policy called "Verpackungsverordnung" in 1991, a legislation to avoid packaging waste. Other countries such as Australia, Belgium and France followed suit before the European Union (EU) introduced the EU Packaging Waste Directive in 1994 which set established collection and recycling targets for EU Member States. It also established requirements for packaging-design across the EU; however, the directive did not require producers to bear waste collection and recycling costs then (Ajani & Kunlere, 2019). Nigeria launched its Extended Producer Responsibility (EPR) programme in 2014 through the National Environmental Standards and Regulations Enforcement Agency (NESREA). The programme makes manufacturers in Nigeria responsible for the management of their post-consumer products.

The policy requires producers to ensure that wastes arising from the use of their products are safely managed through effective monitoring of the entire lifecycle of their products. Individuals and organisations who buy or use products that are at the end of the useful life cycle are also required to safely dispose of them through legal and appropriate means (e.g., use collection centres managed by accredited collectors). The guidelines permit producers who are unable to effectively manage their products' end-of-life wastes to utilise third-party agents (network of collectors and dismantlers and recyclers known as Producer Responsibility Organizations – PROs) to help them oversee the process (Ajani & Kunlere, 2019). This initiative is considered more extensive than the recently proposed plastic legislation as it required businesses with significant waste outputs to have an EPR plan and such a plan must align with the National Environmental Regulations applied to the sector the company belongs to before it is approved by NESREA.

The programme is considered a more comprehensive model that seeks to optimise the benefits of recycling (Woggsborg & Schröder, 2018), but the first fundamental issue when discussing the circular plastic economy in Africa is that, like most other initiatives, this programme is not plastic waste focussed, even though it covered packaging materials such as aluminium, glass, metals, paper and plastics. Secondly, while it has the implication of affecting product design, it offers little in the area of reusability. In general, reviews of the programme in Nigeria also show that the initiative remains largely unknown, and it is often misconstrued by the public, misapplied and underutilised by businesses while its implementation across Nigeria is recorded to continue facing various challenges (Ajani & Kunlere, 2019).

Reviews of this initiative also show that there is poor public participation because many citizens do not understand the benefits nor their roles in the implementation (Ajani & Kunlere, 2019). The initiative also has poor enforcement mechanisms as defaulting companies can easily evade sanctions. There is also a case of insufficient collection centres, and its implementation has been limited to mostly large cities of Lagos, Abuja and Port-Harcourt. These reasons amongst others coupled with poor funding for the implementation of the policy have weakened the likely effectiveness of plastic waste management. We will however consider how digital tools may come to play in solving some of these problems.

4 Descriptive Statistics and Discussions

We utilise both secondary data source (Chatham Database) and primary data source (DITCh Plastic Survey) in this chapter. Our earlier analysis of policies across Africa made use of the secondary data. This segment utilises the primary data. The DITCh data contained responses from 33 entities (17 organisations, 16 other stakeholders that are not digital innovators) and over 1,500 households surveyed (using trained field officers) across five countries in Africa (Kenya, Namibia, Nigeria, Rwanda and Zambia). The survey assesses the level of technological readiness and different digital tools adopted for accelerating the transition to a circular economy in Sub-Saharan Africa. The insight from this segment is used to align existing policy with public perception as a step in suggesting the role digital technologies can play in alleviating these challenges.

We investigate the efficiency of waste management on the continent and about 24% of respondents in the survey clearly see the waste management on the continent as ineffective while about 43% see the waste management on the continent to be either effective or very effective. About 58% believe that the government is not doing enough. This fact further confirms the earlier assertion that there is little evidence on policy success rates (Kweku & Johanna, 2020).

To assess the accuracy of public perception regarding the effectiveness of existing policies, we review the publics' awareness of laws on plastic management. About 82% of respondents are unaware of laws on plastic waste. This is a major insight from the survey as lack of awareness stands out as one of the reasons plastic waste policies remain ineffective in Africa. We then assess the law enforcement confidence of those who claim to be aware of plastic waste laws in their country and found that about 50% of them do not have confidence in the law enforcement. This is considered a reflection of in-built weaknesses in the policies and lack of faith in enforcement agencies. This outcome aligns with Ajani and Kunlere (2019) assertion that the most comprehensive initiative in Nigeria (for instance) remains largely misconstrued by the public, misapplied and underutilised by businesses (Ajani & Kunlere, 2019).

To facilitate an informed policy, we inquire about how some factors may encourage respondents to engage in sustainable plastic waste management. Over 90% of respondents (62.72% – very useful and 28.49% useful) are of the view that political instruments such as legislation will encourage them to engage in sustainable plastic waste management. Assessing how likely plastic waste tax may work, only about 11% are very confident on plastic waste tax and an additional 30% just confident. About 33% are not confident that the imposition of plastic waste tax can influence their waste management habits. This brings to fore the limit of prohibitive tax or use of penalties to drive plastic sustainable behaviour in Africa. This is closely related to our conclusion from text analysis of policy descriptions across Africa.

On the use of economic payments or incentives, about 85% see this as a viable option that can encourage them to engage in sustainable plastic waste management while less than 5% found economic payment as not useful at all. This brings to fore a policy niche often missing in African efforts despite evidence from countries such as India that financial incentives could be effective. Similarly, about 78% of respondents are of the view that they will be encouraged to sustainably manage their plastics waste if they see their friends and family doing it (social influence) while only less than 5% consider it as a not useful approach. This insight provides opportunity to further investigate how social ties and connectedness can be used to drive plastic circular economy in Africa. There are recent evidence that social connectedness can be used to foster positive adjustment behaviour (Turki et al., 2018).

We find that better publicity and awareness will likely be a good strategy to achieve better plastic waste management as over 72% (27.48% – very useful and 45.32% – useful) are of the view that awareness of environmental risks or dangers associated with plastic waste will encourage them to act sustainably. We also find that fair pricing of bottles will be an effective approach as only less than 3% of respondents found it not useful at all and 48.76% of respondents are willing to take their plastic wastes to a collection centre.

We do recognise that every aspect of digital technology (cloud, big data, IoT, blockchain, AI, robotics, GIS, ARvr, websites, 5G, MobileApp, amongst others) has a role to play in managing plastic waste pollution in Africa, but we focus more on mobile solution and websites because of the insight from our data. Only about 25% gave adverse responses to understanding of mobile applications (13.9% never heard of it before and 11.39% are poor at using it), 67.37% are above good (good and excellent) while 7.34% were neutral. Respondents' understanding of website technology is similar to mobile applications. On the other hand, 67.83% gave adverse responses to understanding of AI while only 4.07% are excellent at the use of it, 69.82% gave adverse responses to understanding of Geographic Information System (GIS) while only 3.09% are excellent at using it, 72.97% gave adverse responses to their understanding of blockchain technology while only 2.12% have excellent understanding of the technology. The trend is similar for robotics, cloud (serverless), ARvr and 5G. The fact also that over 74% of respondents claim to have smartphones gives us a further justification to focus on mobile solutions. These results incline us more towards digital-technology-driven solution for Africa.

Presented in Table 14.5 is the respondents ranking of barriers to their adoption of digital tools/technology in plastic waste management (beginning with 1 as the

TABLE 14.5 Summary statistics of respondents ranking of their barriers to adopt digital tools

	Between 1 and 3	Between 4 and 5
Technical barrier	70.10%	29.90%
Economic barrier	74.13%	25.87%
Political barrier	74.46%	25.54%
Socio-cultural barrier	71.01%	28.99%

most significant barrier and 5 as the least significant). The overall image is a high barrier to digital tools adoption.

This sub-segment presents general discussion following insights from the descriptive statistics. It highlights major challenges that have been responsible for ineffective or weak plastic waste policies in Africa and proposes practical ways digital technology can mitigate some of the challenges.

Exclusion of the informal sector – The starting point when we discuss transitioning to a circular economy is always effective waste management. In the case of Africa, the conventional framework for recycling and waste management will lead to suboptimal outcomes if the activities of the informal recycling sector are not brought into the equation (Wilson et al., 2006). An effective policy for Africa will need to integrate plastic waste recycling activities of waste pickers and scavengers (informal recycling sector) into the national plan. Findings show that waste scavengers can be formally integrated into the recycling process. This has been demonstrated to be economically viable (Adeyemi et al., 2001). In that vein, similar technology used in AgriTech can achieve this outcome. Digital technologies have enabled crowd farming in Africa such as Farm Crowdy in Lagos, Thrive Agric in Ghana, and Complete farmer in Ghana and same can be modified to suit informal waste collectors in Africa. It is an aggregation platform that can serve any other sector. Good evidence is the activity of "Mr Green Africa" which is integrating informal waste collectors into the recycling cycle in Kenya.

Awareness problem – (EdTech comes to play). Solutions similar to ones used in EdTech can help in improving masses' awareness of plastic waste policies and in educating them in plastics circular economy. Experience from Fintech shows financial literacy can be enhanced using mobile- and web-based educational platforms. This EdTech can be adopted and extended to educate on plastics wastes' impacts and increase awareness of existing policies. When it comes to educating the masses, Takacycle (Tanzania) is one example that uses waste collection and recycling infrastructure to teach and incentivise people on capturing values from their waste while OkwuEco (Nigeria) is using image recognition to educate households about recycling and linking them to waste merchants.

Enforcement problem (Traceability + Blockchain) – Enforcement of any law becomes almost impossible under asymmetric information (incomplete information). Traceability is however possible with digital tools of instilling barcodes that link every plastic packaging to its manufacturer. This enables monitoring and to appropriately enforce penalties for plastic packaging not properly reused or recycled. Similar tools have been used in tracing and confirming the genuineness of drugs in Africa (Kenya and Nigeria). The simple tool will also facilitate a recycling economy where scavengers are paid for recovered plastic and the subject plastic manufacturer is debited for the recovery activities. This particular tool will also tackle one of the main problems often cited in literature as a challenge to a plastic circular economy, i.e., lack of information on plastic usage.

We also find a more recent application of blockchain technology for smart contracts as an important tool to improve plastic waste policy in Africa. Chidepatil et al., (2020) in their projects (using blockchain smart contracts, AI and multi-sensor data-fusion) presented efforts at segregating plastics based on the plastics' types. This is claimed to be able to efficiently segregate commingled plastics and can result in all actors (segregators, recyclers and manufactures) being able to share data, plan their supply chain, execute purchase orders and further increase the use of recycled plastic feedstock (Chidepatil et al., 2020).

Product redesign which includes reviewing inputs in production and final packaging of products also has a role in transforming Africa to a plastic circular economy but while that will pivot around engineering reviews, big data from digital technology makes engineering redesign easier and cheaper than before. Chidepatil et al. (2020) address how they are able to help manufacturers get reliable information about the availability, quantity and quality of recycled feedstock using advanced blockchain and AI technologies. This was achieved by calibrating and deriving different grades for different recycled polymers. Manufacturers will then be able to assess the suitability of recycled polymers for various applications (Chidepatil et al., 2020).

The combined utilisation of big data, social media data and machine learning (AI) will help to leverage on the social factor (as observed from our survey) to enhance sustainability habits of Africans. The Social Connectedness Index (Facebook and WhatsApp) presents massive data opportunities to utilise network and social connectedness across nations to influence plastic circular economy adoption while data from Twitter can be used to model strength of a network at the individual levels, thereby providing optimal policy targets. Machine learning can be used to demographically classify individuals on social networks (Ajala et al., 2021), such that plastics control initiatives can be appropriately channelled to key actors.

Big data and mobile applications can facilitate a new redistribution model such that products are used to their full potential as users can co-use instead of owning them personally. This can equally originate from a firm managing physical flow of resources better by making use of big data analytics to assess customers' consumption patterns, behaviour to forecast demand. Wireless intelligent technology can also be integrated into the production line.

Advising on policies itself, emphasis should be placed on internalisation of external costs, where companies that control for emissions and pollution are rewarded. Considering the weakness of a linear economy also, the principle of stewardship should be underscored instead of ownership and its right to destroy. Similarly, policies should be extended beyond punitive laws to the use of economic instruments such as the use of incentives and taxes. A business approach justification can be made from eTrash2cash in Nigeria which is already using web, mobile apps and SMS to exchange wastes for direct cash. Economic incentives currently lacking on the continent's regulation can leverage these technologies.

This is similar to what Eco-Post is doing in Kenya. Fundamental change from outright ownership to leasing will increase reusability (a case of Michelin model). This will promote the product as a service.

Generally, we see a need for structural changes to current policies to experience better adoption of digital innovations for circular plastic economy (Berg et al., 2018). We are of the view that digital innovations can be used to create a well-informed cohort of innovators to promote diffusion of circular plastic economy (Kolade et al., 2022). It can also be used to build a more collaborative multi-sectoral community that can advance plastic circular economy in Africa. Digital technologies can generate economies of scale for circular plastic economy stakeholders by connecting stakeholders from a wide range of backgrounds, sectors and countries across Africa. It can also help to create markets for recycled parts (Oyinlola et al., 2022). In the area of policy, it can aid in implementing EPR regulations and in addressing regulatory barriers.

As part of our contribution to plastic circular economy in Africa, we re-echo World Economic Forum assertion that governments should set up political, legislative and economic frameworks that can incentivise profitable circular economies. We then specifically recommend that African countries should formulate policies facilitating digital backbone at national levels (World Economic Forum, 2021b). This is expected to create competing digital circular business models. This will enable small and medium enterprises (SMEs) to participate in circular economy against the current trend where large multinational companies are leading the drive. This backbone will enable interoperability of many-to-many against the one-to-one interoperability often experienced in the linear economy. We are of the view that it will enable SMEs to scale their innovations. Its potential to reduce cost and risk when it comes to circular economy will help the circular plastic economy. Such a backbone will allow data sharing and standardisation. We find confidence in this suggestion drawing inference from Mojaloop (an open-source software), designed to provide a reference model for payment interoperability. This has already been adopted by some national governments (e.g., Rwanda) with the hope that the interoperability will help in overcoming barriers that have slowed the spread of digital financial services across Africa.

4.1 Evidence of Emerging Digital Tools in Africa Plastic Economy Ecosystem

While policy has a significant part to play in driving the plastic circular economy, we are witnessing a massive role played by digital innovation in improving material efficiency. Primary discourse of circular economy revolves around large corporations because of their perceived capabilities to both conceptualise and lead transformation to circular economy (Schröder et al., 2019). However, we are of the view that African reality such as, the presence of large informal sector, significant role of government in the economy and the profit-maximisation inclination of large businesses, will demand a slightly different approach to scaling the circular economy in Africa.

We briefly present evidence of the possibility of digital technology, playing a role in addressing plastic waste problems in Africa. We showcase some selected tech start-up ideas focusing on plastic waste problems in Africa (Table 14.6). We map their business ideas to different areas of the plastic circular economy action areas. The three broad action areas often discussed in creating a plastic circular economy are "eliminate, innovate and circulate". The actions are on eliminating all unnecessary plastics, ensuring reusability, recyclability and composability of plastics and to continuously circulate plastic to keep them out of the environment.

5 Summary and Conclusion

The concept of "circular economy" presents a beneficial loop of continuous material recycling without the adverse effects of new production on the environment; however, recycling is only one and of a lower order in the hierarchy of reducing plastic waste impacts (Allwood, 2014). Other policies that reduce demand and increase re-use of products are strategies with equally great potentials to transform Africa into a circular economy. While large multinational companies such as Michelin might be leading adopters of the circular plastic economy globally, the African experience will need to rely on tech start-ups in Africa. The result from our survey shows how difficult it may be for companies in Africa to fully adopt the multinationals model, but tech start-ups have the opportunity to capitalise on this and redefine the business model within the continent. They have access to enabling digital technologies to scale this new business model and they can develop capacities to create circular advantages from product design to production and profitable regeneration.

Many countries have adopted national and international policies targeting plastic pollution, but substantial numbers of people, firms and organisations will still need to alter their existing behaviours if global plastic pollution is to be curbed. Evidence has shown that education and persuasion alone are insufficient to achieve this outcome, therefore making government policies imperative (Kinzig et al., 2013). As Stahel (2016) suggested, there is the need for governments and regulators to adopt policies that will promote a circular economy at the industry level, including the use of taxation. Likewise, innovations to pave the way for further advancement in splitting up molecules to re-cycle atoms should be supported by the government.

We do recognise that a circular economy will be beneficial to all stakeholders, but many organisations in Africa are not currently built to capitalise on circular advantage. The transformation from a linear to a circular economy will require not only an environmental but also a social and economic restructuring of

TABLE 14.6 Mapping of selected tech start-ups in Africa to plastic circular economy areas

	Eliminating unnecessary plastics	Recycling used plastics	Ensuring reusability of plastics
Recycle Bot (Zambia)		Mobile device used across the whole value chain	
Africa– Waste – Veolia (Côte d'Ivoire)	Users are able to indicate the amount of waste they would like to remove		
Capture Solutions (Nigeria)	Digitisation of processes (IoT Devices).		Geotagging of activities for material traceability. Community based training
Chanja Datti (Nigeria)		Online based recycling company (consumers are rewarded for recycling)	
Coliba (Côte d'Ivoire)	Platform to request pick up of plastic wastes		
Ecofuture (Nigeria)	Collects recyclable plastic wastes using mobile app and SMS		
Kaltani (Nigeria)	Collect and sort plastic	Recycle plastic waste	Wash plastic waste
Recuplast (Senegal)	Website based collection of plastic waste		
Salubata (Nigeria)			Online store selling modular shoes made from recycled plastic
SOSO Care (Nigeria)		Provide health insurance where recyclables are premium	
Dispose Green (Ghana)		Apps that connect people to a wide network of waste collectors	
EasyWaste (Ghana)		Operate recycling centres	
Takacycle (Tanzania)			Educating people on how to capture value from their waste.
Wastezon (Rwanda)	App/Website based waste pick up		Selling of wastes using an app

WasteBazaar (Nigeria), RecycleGarb (Nigeria), GreenHill Recycling (Nigeria), MIRA (Ghana), ComeRecycle (Nigeria), Eazy Waste (Ghana), Zonku Technology (Uganda), Yo–Waste (Uganda), Wrapp (South Africa), Virdismart (Kenya), Vicfold recyclers (Nigeria), Scrapays (Nigeria), Reveal Uno (Ghana).

Note: Data extracted from DITCh Innovation Aggregator Website – (Last assessed on October 31, 2021)

production and consumption patterns. Our first proposition was how to integrate the informal sector into the circular economy? Digital tools can be used to bring together the activities of the informal recycling sector. To increase cooperation among nations, data banks and blockchain technology that facilitate traceability of plastic wastes to their source producers/countries will significantly help not only in formulating better plastics waste policy, it will enhance enforcement.

Our conclusions align with da Costa's (2018) position that four major reasons can explain why current efforts at transforming the plastic economy have yielded limited success. That there is insufficient regulatory scope while for existing regulations, there is lack of implementation and enforcement. Also, there is insufficient states' participation in regional initiatives (poor international cooperation) coupled with inexistence of sufficient data on the prevalence of marine plastic waste in the environment.

We reiterated that some start-ups within the African tech ecosystem are already incubating businesses that can improve plastic waste policies on the continent such as EasyWaste which has been serving as a data hub for reporting collection of waste and recycling data. They also assert to be helping policymakers formulate good waste management policy and bring plastic scavengers from the informal sector. WeCyclers is also using an app to store the number of collected recyclables from various locations while Virdismart uses automated waste collection and management, making use of a Smart Bin that rewards customers.

This chapter has added to literature in two ways. Firstly, it reviewed initiatives in Africa that were intended to affect circular plastic economy on the continent and undertook a country plastic circular economy review of Nigeria (Nigeria Plastic Regulation Bills and the Extended Producers Responsibility Programme), thereby adding to the limited literature on plastic policies in Africa. Secondly, it highlighted digital technological tools and how the tools can be used to enhance Circular Economy Policy effectiveness in Africa and presented some current efforts been made by start-ups in Africa to attend to plastic waste challenges.

Appendix 1

Country	Policy Count
Nigeria	14
South Africa	10
Egypt	9
Mauritius	9
Ghana	8
Tanzania	8
Tunisia	8

Country	Policy Count
Uganda	8
Cameroon	7
Rwanda	7
Seychelles	7
Algeria	6
Benin	6
Cape Verde	6
Madagascar	6
Mozambique	6
Namibia	6
Togo	6
Zimbabwe	6
Burkina Faso	5
Kenya	5
Malawi	5
Mali	5
Zambia	5
Angola	4
Botswana	4
Burundi	4
Côte d'Ivoire	4
Democratic Republic of the Congo	4
Djibouti	4
Gambia	4
Niger	4
São Tomé and Príncipe	4
Senegal	4
Ethiopia	3
Gabon	3
Mauritania	3
Morocco	3
Comoros	2
Guinea	2
Libya	2
Republic of the Congo	2
Sierra Leone	2
Somalia	2
Sudan	2
Central African Republic	1
Chad	1
Eritrea	1
Eswatini (formerly Swaziland)	1
Guinea-Bissau	1
Lesotho	1
Liberia	1

References

Adeyemi, A. S., Olorunfemi, J. F., & Adewoye, T. O. (2001). Waste scavenging in Third World cities: A case study in Ilorin, Nigeria. *Environmentalist, 21*(2), 93–96.

Ajala, O., Ejiogu, A., & Lawal, A. (2021). Understanding public sentiment in relation to the African Continental Free Trade Agreement. *Insight on Africa,* 09750878211012884.

Ajani, A., & Kunlere, I. (2019). Implementation of the Extended Producer Responsibility (EPR) Policy in Nigeria: Towards sustainable business practice. *Nigerian Journal of Environment and Health, 2*(1), 44–56.

Alam, K., & Mahal, A. (2014). Economic impacts of health shocks on households in low and middle income countries: A review of the literature. *Globalization and Health, 10*(1), 1–18.

Allwood, J. M. (2014). Squaring the circular economy: The role of recycling within a hierarchy of material management strategies. *Handbook of recycling* (pp. 445–477). Elsevier.

Austin, J., Hatfield, D. B., Grindle, A. C., & Bailey, J. S. (1993). Increasing recycling in office environments: The effects of specific, informative cues. *Journal of Applied Behavior Analysis, 26*(2), 247–253.

Ayeleru, O. O., Dlova, S., Akinribide, O. J., Ntuli, F., Kupolati, W. K., Marina, P. F., Blencowe, A., & Olubambi, P. A. (2020). Challenges of plastic waste generation and management in sub-Saharan Africa: A review. *Waste Management, 110,* 24–42.

Bateman, S. (2007). UNCLOS and its limitations as the foundation for a regional maritime security regime. *The Korean Journal of Defense Analysis, 19*(3), 27–56.

Berg, A., Antikainen, R., Hartikainen, E., Kauppi, S., Kautto, P., Lazarevic, D., Piesik, S., & Saikku, L. (2018). Circular economy for sustainable development. Finnish Environment Institute (ISBN: 978-952-11-4970-2)

Boulding, K. E. (2011). *The economics of the coming spaceship earth.* H. Jarrett (ed.). *Environmental quality in a growing economy* (s. 3–14). New York: RFF Press. https://doi.org/10.4324/9781315064147

Cagno, E., Neri, A., Negri, M., Bassani, C. A., & Lampertico, T. (2021). The role of digital technologies in operationalizing the circular economy transition: A systematic literature review. *Applied Sciences, 11*(8), 3328.

Chatham House. (2021). *circulareconomy.earth.* Retrieved November 1, 2022, from https://circulareconomy.earth/

Chidepatil, A., Bindra, P., Kulkarni, D., Qazi, M., Kshirsagar, M., & Sankaran, K. (2020). From trash to cash: How blockchain and multi-sensor-driven artificial intelligence can transform circular economy of plastic waste? *Administrative Sciences, 10*(2), 23.

Chow, C., So, W. W., Cheung, T., & Yeung, S. D. (2017). Plastic waste problem and education for plastic waste management. *Emerging practices in scholarship of learning and teaching in a digital era* (pp. 125–140). Springer.

da Costa, J. P. (2018). Micro-and nanoplastics in the environment: research and policymaking. *Current Opinion in Environmental Science & Health, 1,* 12–16.

da Costa, J. P., Mouneyrac, C., Costa, M., Duarte, A. C., & Rocha-Santos, T. (2020). The role of legislation, regulatory initiatives and guidelines on the control of plastic pollution. *Frontiers in Environmental Science, 8,* 104.

Desmond, P., & Asamba, M. (2019). Accelerating the transition to a circular economy in Africa: Case studies from Kenya and South Africa. *The Circular Economy and the Global South* (pp. 152–172). Routledge.

Dijkstra, H., van Beukering, P., & Brouwer, R. (2020). Business models and sustainable plastic management: A systematic review of the literature. *Journal of Cleaner Production, 258*, 120967.

Gagain, M. (2012). Climate change, sea level rise, and artificial islands: Saving the Maldives' statehood and maritime claims through the constitution of the oceans. *Colorado Journal of International Environmental Law and Policy, 23*, 77.

Hornik, K., & Grün, B. (2011). Topicmodels: An R package for fitting topic models. *Journal of Statistical Software, 40*(13), 1–30.

Kaza, S., Yao, L., Bhada-Tata, P., & Van Woerden, F. (2018). *What a waste 2.0: a global snapshot of solid waste management to 2050*. World Bank Publications.

Kinzig, A. P., Ehrlich, P. R., Alston, L. J., Arrow, K., Barrett, S., Buchman, T. G., Daily, G. C., Levin, B., Levin, S., & Oppenheimer, M. (2013). Social norms and global environmental challenges: The complex interaction of behaviors, values, and policy. *Bioscience, 63*(3), 164–175.

Kolade, O., Odumuyiwa, V., Abolfathi, S., Schröder, P., Wakunuma, K., Akanmu, I., Whitehead, T., Tijani, B., & Oyinlola, M. (2022). Technology acceptance and readiness of stakeholders for transitioning to a circular plastic economy in Africa. *Technological Forecasting and Social Change, 183*, 121954.

Kweku, A., & Johanna, T. (2020). *Policy approaches for accelerating the circular economy in Africa*. Retrieved November 6, 2022, from https://circulareconomy.earth/publicati ons/accelerating-the-circular-economy-transition-in-africa-policy-challenges-and-opportunities

Lacy, P., & Rutqvist, J. (2016). *Waste to wealth: The circular economy advantage*. Springer.

Luo, Y., Douglas, J., Pahl, S., & Zhao, J. (2022). Reducing plastic waste by visualizing marine consequences. *Environment and Behavior, 54*(4), 809–832.

MacArthur, D. E., Waughray, D., & Stuchtey, M. R. (2016). The new plastics economy, rethinking the future of plastics. Paper presented at the *World Economic Forum,*

Oyinlola, M., Schröder, P., Whitehead, T., Kolade, O., Wakunuma, K., Sharifi, S., Rawn, B., Odumuyiwa, V., Lendelvo, S., & Brighty, G. (2022). Digital innovations for transitioning to circular plastic value chains in Africa. *Africa Journal of Management, 8*(1), 83–108.

Rochman, C. M., Browne, M. A., Halpern, B. S., Hentschel, B. T., Hoh, E., Karapanagioti, H. K., Rios-Mendoza, L. M., Takada, H., Teh, S., & Thompson, R. C. (2013). Classify plastic waste as hazardous. *Nature, 494*(7436), 169–171.

Schmidt, C., Krauth, T., & Wagner, S. (2017). Export of plastic debris by rivers into the sea. *Environmental Science & Technology, 51*(21), 12246–12253.

Schröder, P., Anantharaman, M., Anggraeni, K., & Foxon, T. J. (2019). *The circular economy and the Global South: sustainable lifestyles and green industrial development*. Routledge.

Stahel, W. R. (2016). The circular economy. *Nature News, 531*(7595), 435.

Syberg, K., Nielsen, M. B., Clausen, L. P. W., van Calster, G., van Wezel, A., Rochman, C., Koelmans, A. A., Cronin, R., Pahl, S., & Hansen, S. F. (2021). Regulation of plastic from a circular economy perspective. *Current Opinion in Green and Sustainable Chemistry,* 100462.

Turki, F. J., Jdaitawi, M., & Sheta, H. (2018). Fostering positive adjustment behaviour: Social connectedness, achievement motivation and emotional-social learning among male and female university students. *Active Learning in Higher Education, 19*(2), 145–158.

Wilson, D. C., Velis, C., & Cheeseman, C. (2006). Role of informal sector recycling in waste management in developing countries. *Habitat International, 30*(4), 797–808.

Woggsborg, A., & Schröder, P. (2018). Nigeria's e-waste management: Extended producer responsibility and informal sector inclusion. *Journal of Waste Resources and Recycling, 1*(1), 102.

World Economic Forum. (2021a). *17 innovations accelerating the transition to a circular economy.* Retrieved November 6, 2022, from www.weforum.org/agenda/2021/04/17-innovations-accelerating-the-transition-to-a-circular-economy

World Economic Forum. (2021b). *Why digitalization is critical to creating a global circular economy.* Retrieved November 6, 2022, from www.weforum.org/agenda/2021/08/digitalization-critical-creating-global-circular-economy/

15

GENDER AND DIGITAL INNOVATION ON CIRCULAR PLASTIC ECONOMY IN AFRICA

Kutoma Wakunuma and Selma Lendelvo

1 Introduction and Background

Digital innovation presents many opportunities for CPE which include the creation of job opportunities in a relatively niche area such as informal waste picking which is an essential source of employment for people: opportunity to develop technical skills as well as an opportunity to reduce the gender disparity gaps that exist when it comes to plastic waste and management. This section will therefore look into what digital innovation means in CPE, discuss gender disparities in a digitally enabled CPE and consider the landscape of digital innovations on CPE in Africa.

1.1 Digital Innovations and the Circular Plastic Economy

Digital innovation is defined as a product, process or business model that is new or requires significant changes and it is enabled by information technologies (Yoo, Hendfridsson & Lyytinen, 2010; Berg & Wilts, 2019). This digital sector is comprised of a wide variety of inventions and emerging use of technologies such as internet of things (IoT), artificial intelligence, mobile technologies, advanced robotic, cloud computing, 3D printing, big data virtual reality, blockchains and autonomous vehicles etc. (Rymarczyk, 2020; European Environment Agency, 2021).

Many sectors, including financial industry, transportation, education and the waste management industry, have been significantly influenced by the digital innovation digital platforms (Yousaf, Radulescu, Sinisi, Serbanescu & Păunescu, 2021). Waste management industry, which forms the basis of the study, has also experienced the application and usage of different technologies to tackle the issue of mismanaged plastic waste, which is growing exponentially (Joshi, Seay

DOI: 10.4324/9781003278443-18

& Banaddab, 2018). Digital technologies are increasingly applied across all areas of waste collection, and certain aspects of collection have been transformed by advanced digitalisation, especially "logistics, the process of organising, scheduling and dispatching tasks, personnel and vehicles, here digital tools offer the potential to enhance the process by storing, processing, analysing and optimising the necessary information" (European Environmental Agency, 2021).

Although plastic waste is a global catastrophe, challenging both developed and developing countries, the magnitude and extent of impacts vary. Low-to-middle-income countries (LMICs) found in Asia, South America and Africa associated with growing economies, urbanisation, resource constraints, limited infrastructure, lack of effective governmental policy and regulations, along with insufficient household education and scarce stakeholder involvement are identified to be fighting triple battles with regards to plastic waste accumulation and management (Joshi, Seay & Banaddab, 2018). However, local governments in low- and middle-income countries are, on average, only collecting half of the waste generated in urban centres (Joshi, Seay & Banaddab, 2018). In response to this challenge, individuals and small communities are being empowered to adapt and invent locally managed innovative circular economy models rather than waiting on central authorities to address the problem (Joshi, Seay & Banaddab, 2018). The long-term objective of these institutions is to achieve a circular plastic economy (CPE). The circular economy approach is focussed on replacing the unsustainable linear economy of produce–consume–discard by promoting reusing, recycling or re-entering products into their manufacturing supply chain (Kaur, Uisan, Ong & Ki Lin, 2017).

Circular economy as an alternative model for plastic waste management has emerged over the past several decades. This model is regarded to be an opportunity to contribute to sustainable development through sustainable plastic production and consumption, thus enabling economic, social and environmental sustainability (Stahel, 2016; Nandi, Sarkis, Hervani & Helms, 2021).

Transitioning to CPE in African countries will require a combination of effort by many actors to work sometimes in coalition, as well as in unstructured form. Africa has also experienced the emergence of international organisations which supports the sharing of good circular practices. For instance, the Circular Economy Club is a global network of circular economy professionals which encourages collaboration to achieve a greater impact of circular practices. Moreover, numerous actors in Africa across different spheres are building on the effort to create a more sustainable future for the continent: some are non-governmental organisations (e.g. The African Circular Economy Network), businesses (both multinational and smaller entrepreneurs), international development agencies (e.g. Tearfund) and international coordinating bodies (e.g. World Economic Forum) (Desmond & Asamba, 2019) and governmental agencies such as the African Circular Economy Alliance (ACEA), whose aim is to encourage advocacy projects through policy research that support high-impact circular economy projects. This initiative includes such countries like South Africa, Rwanda, Nigeria among others as well as the DITCh Plastic Network [born out of the DITch project – Ditchplastic.org].

The use of emerging technologies such as IoT, artificial intelligence and mobile technologies are significant in the realisation of innovations among both men and women. For decades, circular activities have provided new and different kinds of skills and jobs (Schmitz, 2015). For example, a digital innovation start-up innovation in Nigeria uses a mobile app to connect private-sector waste collectors with sources of segregated waste (including plastic) from households and businesses. Moreover, digital technology was found to improve traceability of plastic, for example, in India, a mobile platform provides urban waste pickers with access to fair market prices, while also allowing the buyers of recycled plastic to trace the source of their materials (GSMA, 2021).

Digital technologies were also found to enhance awareness and education tools to promote awareness on health risks associated with mismanaged waste. For instance, mobile technologies are being highly used to influence positive behaviour change through information sharing, incentives or gamification, to encourage and educate consumers on the choices they make and the impact and reward of their choices (GSMA, 2021).

It is being widely recognised that the emergence and expansion of digital innovations and technologies are catalysts and drivers of social and economic development (Oliveira, Oliver & Ramalhinho, 2020). Moreover, Park and Choi, (2019) asserted that the continuous advancement of digital innovation and technologies has the capacity to improve national economies and social and industrial development. Similarly, a study by Khan, Khan and Aftab (2015) showed how digital innovation highly contributed to the achievement of long-term economic growth, reducing unemployment as well as contributing to the improvement in the quality of life of the people. Hence, CPE and digital innovations are a promising alternative for improving household and community-level economies (Schröder, Lemille & Desmond, 2020).

Furthermore, digital innovation has transformed the global traditional entrepreneurial and business models in different industries into a digitalised economy (Bukht & Heeks, 2018; Barefoot, Curtis, Jolliff, Nicholson & Omohundro, 2018). The adoption of digital technologies based on the network connectivity is accelerating and fostering easier and efficient exchange of information, easier collaboration and networking within and across national borders (Park & Choi, 2019; Yousaf, Radulescu, Sinisi, Serbanescu & Păunescu, 2021).

1.2 Gendered Hazards and Risks in the Circular Plastic Economy

Digital technologies impact all the social life areas. Hence, the shift towards a digital world has had cross cutting effects on society that extend beyond the digital technology context (Balcerzak & Pietrzak, 2017; Yousaf et al., 2021). The recognition of the importance of digital innovation in economic development has also led to debates and questions around gender inequalities due to the fact that the digital world has mainly been male-dominated. These similar patterns tend to recur in the CPE sector which is heavily male-dominated, thus exacerbating persistent gender disparities. Gender disparity was found to be visible in policies, legislation and strategies which are heavily skewed towards men. In Ghana, it was shown that there was unequal representation of women among decision makers in regulatory institutions, therefore limiting the participation of women when policy decisions are being made (Global Plastic Action Partnership, 2021; OECD, 2020).

The sector that tends to favour more women is the informal economy, where women work predominantly as main waste pickers and employed in recycling companies as street sweepers, washers and sorters (Ocean Conservancy, 2019; OECD, 2020; Global Plastic Action Partnership, 2021). In Africa and Asia, it was revealed that some waste management institutions favoured higher participation of women in informal waste collection from landfills and dumpsites, as well as typically employing them to perform tasks of packing, washing and sorting waste. Tasks deemed to be more labour intensive such as collecting, lifting and loading of recyclable materials and are generally considered more suitable for men (Ocean Conservancy, 2019). However, as packing, washing and sorting are also labour intensive, the clear distinction appears to be more related to age-old social norms where certain tasks, particularly those done at home, such as washing, are assumed to be for women while those related to lifting heavy objects, for instance, are assumed to be for men.

1.2.1 Health Hazards

Plastic products are noted to contain harmful chemicals. Some of these chemicals are enhanced through the re-washing and re-use of containers (Ayeleru et al., 2020). Hence, women face greater health and safety risks as a result of waste handling, picking and manual recycling of products and/or working in unhygienic environments. For instance, in Ghana, a report showed that during plastic waste collection at dumpsites, some women owned a metallic tool or stick for picking and others wore gloves, while the majority used their bare hands for picking the plastic waste (Global Plastic Action Partnership, 2021). Furthermore, a study conducted in India, Vietnam, Indonesia and Philippines revealed that there were high incidences of respiratory illnesses (including lung cancer) among waste pickers due to exposure to toxic/hazardous plastic waste and other unhygienic materials. This was exacerbated by lack of health insurance coverage and limited access to adequate health care. Such risks affected their ability to participate in economic opportunities through sales of their recyclables collected as well as in managing their households (Ocean Conservancy, 2019).

1.2.2   Physical Risk and Safety

Risk and safety are also among the challenges identified by women involved in waste value chain. The waste sector is an unregulated sector; hence, when women walk on foot to dumping sites, which are in most cases located in alleyways and poorly maintained areas of the city, they get exposed to crime and risky situations (Ocean Conservancy, 2019). In addition, as most of the waste is dumped at night or in the early hours when it is still dark, women's safety is compromised. However, they are forced to compete with their male counterparts who do not have the same safety and security concerns during those hours (Ocean Conservancy, 2019).

1.2.3   Cultural Norms

The CPE is also identified to reinforce an imbalance of roles on plastic management due to cultural norms. For instance, plastic picking is seen as a dirty job and therefore left to women who are not in decision-making roles. Studies have shown the implication of stigma associated with waste collection and picking. In Indonesia and African countries like Ghana, waste collection is associated with the stigma of being labelled names and not being respected in the society. This has hindered women participation in the occupation of waste collection. Particularly in Indonesia, women narrated that even if faced with extreme poverty, options of begging in the street are more sensible as compared to waste collection (Ocean Conservancy, 2019; Godfrey et al., 2018).

1.2.4   Income and Earnings

The informal recycling economics also reveals some gender variations in minimal returns and earning for women in CPE. Studies have shown that men are able to negotiate better prices for their materials as well as trade in larger volumes than women. This results in men earning more for their labour. These limited opportunities can exacerbate the never-ending cycle of poverty (Ocean Conservancy, 2019; Godfrey et al., 2018; OECD, 2020).

Moreover, capacity building and access to facilities and resources such as empowerment, credit schemes, training and equipment to enhance women participation in digital innovations and CPE is minimal. Even in the informal economy, it has been found that itinerant women pickers predominantly use bags or sacks to collect and transport their collected recyclables; meanwhile, their male counterparts have access or ownership to equipment such as the pushcarts and tricycles. Lack of equipment to assist women involved in waste picking to easily transport their recyclables often limits the choice of recyclables to pick and trade, as they have to travel longer distances (Ocean Conservancy, 2019; Godfrey et al., 2018; OECD, 2020). Hence, if circular economy frameworks and strategies do not integrate gender equality goals, the digitalisation and modernisation of waste management, which is generally capital and technology intensive, will reduce opportunities for less qualified labour, particularly women (OECD, 2020).

1.3 Landscape of Digital Innovations on CPE in Africa

Over the past two decades, Africa's digital economy experienced a wave of technological innovations, although the digital revolution is not continentally uniform (Mourdoukoutas, 2017). Countries termed as the KINGS (Kenya, Ivory Coast, Nigeria, Ghana and South Africa) are known to be at the forefront in terms of diverse technological innovations supported by both public and private sectors. Additionally, countries such as Rwanda, Egypt, Tunisia, Mauritius and Morocco are also known for their growth in digital economies (Portulans Institute, 2021). Although regional disparities exist in terms of digital innovations across the African continent, the realisation on the importance of digital innovation in economic development is growing rapidly and generating interest among other African countries as captured by the World Bank's Digital Economic for Africa Country Diagnostics Status Report (2021).

Increasing trends in digital innovations which began with the spread of the internet, mobile-based technologies, including artificial intelligence (AI), IoT, 3D printing and Blockchain, are now the basis for the growth and use of these and other new and emerging technologies in CPE. These technologies are being implemented to close the gaps in CPE to allow for smart collection and recycling of plastic wastes (Wilson et al., 2021).

Using digital technologies, different institutions in many countries are building extensive networks to improve the efficiency and effectiveness of plastic supply chain from waste collectors, aggregators, transporters and processors from across the informal and formal sectors in the effort to collect and recycle massive amounts of plastic waste that might otherwise be mismanaged and discarded irresponsibly (Wilson et al., 2021).

Among the different technological innovations, Africa is noted to be experiencing an increasing trend of digital innovations such as mobile apps as a result of the ubiquitous nature of mobile phones on the continent. The Global Innovation Index indicates that from 2014 to 2018, over 130 million users were connected to mobile internet in Sub-Saharan Africa, although access and usage of Information and Communication Technologies, online public services and e-participation are noted to be higher in North Africa than in Sub-Saharan Africa (Portulans Institute, 2021).

The Global System for Mobile Communications Association detailed that Africa remains the fastest-growing mobile phone market in the world, with estimated coverage of 725 million smartphone users by 2020 (Mourdoukoutas, 2017). Moreover, the United Nations agency for information and communication technologies reported that about 80.8% of Africans owned a mobile phone in

2016. However, women in Africa, particularly in Sub-Saharan Africa are "15% less likely to own a mobile phone than men" (Shanahan & Rowntree, 2019).

Despite this gender disparity, in addition to using mobile apps for communication purposes, mobile apps are being enhanced and equipped to address a community's specific needs (Mourdoukoutas, 2017), including opportunities to participate in waste management and CPE. Thus, different start-ups are leveraging on such opportunities.

Digital technologies have a critical role to play in improving citizen engagement in plastic waste management. Connections among actors are equally essential to facilitate learning, technology adoption and the development of new technologies. This requires networking and collaboration capabilities among all actors (United Nations, 2018) to enable acquisition of new knowledge, skills, awareness of different technologies and networking. This is an essential requirement for technology transfer, which is a complement to, not a substitute for, efforts to build endogenous innovation potential (United Nations, 2018).

In proceeding to the method adopted for the study, a recap of the objectives of the study are listed below:

(i) To identify opportunities to make the CPE sector inclusive and gender responsive
(ii) Interrogate the gender inequalities in CPE
(iii) To understand how digital innovation can reduce these disparities to provide opportunities for both men and women to participate and benefit equally

2   Method

Small-scale CPE practices at individual and community level have been documented in different geographic locations around the world, particularly in China, India and South America. In contrast, African case studies have limited documentation through academic research (Desmond & Asamba, 2019). The study is assessing gender and digital innovations in transition to CPE, using the qualitative method. Data was drawn from focus-group discussions with different sectors across the African continent from Namibia, Rwanda, Nigeria and follow-up interviews from Zambia, Nigeria and Uganda. The use of focus group discussion methodology has a huge benefit from the group setting since it gives understanding into social interactions, and the discussions from the group(s) reflect the overlaps in perceptions and knowledge, including experiences, better (O. Nyumba et al., 2017). A total of 60 respondents participated in focus group discussions, inclusive of start-ups/innovators, investors, civil society, government and parastatals and academia.

A gender analysis framework was used to identify and capture key issues on gender and plastics along the value chain.

LEVELS OF EMPOWERMENT	DESCRIPTION
CONTROL ⬆	Women and men have equal control over factors of production and distribution of benefits, without dominance or subordination.
PARTICIPATION ⬆	Women have equal participation in decision-making in all programs and policies.
CONSCIENTIZATION ⬆	Women believe that gender roles can be changed and gender equality is possible.
ACCESS ⬆	Women gain access to resources such as land, labour, credit, training, marketing facilities, public services, and benefits in an equal basis with men. Reforms of law and practice may be prerequisite for such access.
WELFARE	Women's material needs, such as food, income and medical care are met.

Adapted from Longwe, 2002

FIGURE 15.1   The levels of empowerment of gender analysis using the Women's Empowerment Framework – Adapted from Longwe, 2002

The analysis of this chapter applied the Women's Empowerment Framework (WEF), which focuses on assessing levels of equalities between men and women while taking into consideration the levels of empowerment (March, Smyth & Mukhopadhya, 1999; Longwe, 2002). This framework has five levels of empowerment; however, for this chapter, only two upper elements were applied, which assessed the positive, neutral or negative impact on the target population (Figure 15.1). This will include the level of participation in understanding equal participation in decision making in different programmes and policies. Secondly, the control element was also analysed to establish control over decision-making processes, predominantly evaluating if women and men have both control over the factors of production and distribution of benefits, without dominance or subordination (Longwe, 2002).

Thematic coding was conducted for the focus group discussions to generate key digital innovation and gender themes, which at the end provided an understanding of the people's participation regarding gender landscape in the CPE. The coding process was also simultaneously considering the levels of equality from the WEF.

TABLE 15.1 Details of the FGD participants from different countries across sectors

Sector	Total FGD participants	Male (%)	Female (%)	Not specified*
Digital innovation starts-ups	18	18	55	27
Investors	12	29	57	14
Civil society	15	33	56	11
Government and parastatals	15	67	33	
Waste management organisations	22	31	38	31
Academia	18	82	22	

* These are participants who did not specify their genders during the FGDs

3 Results

3.1 Women Participation in Policy Regulation and in Decision Making

As seen from Table 15.1, government ministries and parastatals constituted more men than women who participated in the FGDs. These are stakeholders from Rwanda, Namibia and Nigeria with oversight responsibility of legal and policy frameworks for environment management, provision of guidelines and standards, to ensure enforcement of all waste, including plastic waste, management activities. The findings indicate that there is still low inclusion of women in decision-making positions related to policy and CPE. Similarly, a study conducted in Ghana found that, the unequal representation of women among decision makers in regulatory institutions limits the participation of women when policy decisions are being made (Global Plastic Action Partnership, 2021).

Stakeholders particularly from Namibia and Rwanda highlighted that, while efforts are being made with regards to implementing policies related to the environment and waste management to be gender inclusive/neutral, there are weaknesses and challenges faced, which needs to be capitalised on to implement policies effectively.

A stakeholder from Namibia remarked *"Many African countries are known in the formulation of good policies and acts, and very poor in the coordination and enforcement of those policies"*.

Meanwhile, a stakeholder from Rwanda was of the view that issues that prevent policy from really being carried out involves aspects such as acceptance and power to enforce regulations, as narrated below:

There is an attempt to incentivize collection of plastic waste. Invitation to invest in alternative packaging is slow to be accepted. Also, there is limited enforcement of regulations at the grass roots level

Government stakeholder, Rwanda, 2020

It is therefore important to have strengthened relations and collaborated efforts with different actors involved in CPE which should include women. The inclusion of women is important in the plastic waste economy in order to help transform a well-balanced and inclusive socio-economic level of the population. Civil society and government sectors can be important sources of information on how plastic waste actually integrates with other problems and for learning how to include stakeholders. This is especially important when it comes to supporting innovation. This will happen by advocating and engaging the community in both urban and rural areas on different opportunities and policy frameworks.

3.2   Digital Innovations Among Waste Management Start-Ups

The focus group discussions also involved starts-ups and revealed that, innovations and start-ups initiated by women are few. This was attributed to a niche competitive area which does not readily allow women to compete on a playing field. Various challenges were identified which included a limited access to education, finance, labour, technology and equipment. Further, it was observed that because technological innovations were often adopted from outside the countries under consideration such as Rwanda, the resulting effect was a limited capacity in technology know-how. In relation to the challenges mentioned above, a study by the Global Plastic Action Partnership (2021) shows that the lack of niche competitive area/industry exacerbated the plastic supply chain industry to be heavily skewed towards men in business ownership and decision making.

Stakeholders raised the prevalence of high attrition of female start-ups in the recycling and waste collection sectors, as opposed to ownership and management of waste collection organisations or innovative companies.

A stakeholder from Nigeria indicated:

> The independent scavengers and recyclers need to be well integrated into a system that works

3.3   Access to Finance

Discussions with investors were also held in order to understand aspects on institutional investing, equity investing and partnerships. Observation from the discussions points to the fact that digital innovation and CPE are not uniform; hence, feasible and CPE country-specific innovation requires fact findings and marketing orientation to enable different prospects for investment. Investors from Namibia were of the view that plastic chain supply opportunities exist, however, that there is not a lot of innovation in CPE in the country as well as lack of fundable private-sector projects. Moreover, investors from Rwanda emphasised that new companies can be enablers of transition to circular economy by exploring niche alternatives such as waste management (material design, optimisation), or transformers (business solutions that change the production and supply systems), before seeking institutional investors. The stakeholders were of the view that lack of financing for alternatives might be partly addressed by new start-up ideas, competitions or incentives.

The number of women accessing financial resources for start-ups was said to be limited. Some stakeholders were of the view that while the male counterparts have ease of access to resources and assets such as information on investments, finances and land, women are often disadvantaged as they lack this crucial information, thus requiring information and awareness creation:

> It was raised that there is lack of starts up and SMEs knowledge and awareness on funding opportunities as well as the establishment of financial feasibility projects; this is exacerbated by weak connections between local business and financial institutions
>
> *Investment stakeholder from Namibia, 2020*

The discussion further acknowledged that incentives and enumeration for those already involved in waste picking is marginal; this is influenced by market demands, competition and fluctuating market value prices. A stakeholder from Rwanda was of the view that low incentives is affecting women participation:

> Incentives for waste collection by waste pickers in the country is rather low or marginal, this has discouraged interest and participation
>
> *Stakeholder from Rwanda*

The lack of incentives can contribute to a lack of interest and participation in the CPE, which by default can result in the prevalence of uncollected plastic waste that is done mainly by women who support production in the value chain by washing, sorting and packaging of plastic waste. Studies such as those conducted in Ghana have documented the decrease of women participation in the labour markets (Baah-Boateng & Twum, 2018).

3.4 *Plastic Waste Chain Stereotype, Roles and Workforce*

Discussions with stakeholders from academia and civil society revealed that there are gender gaps in the plastics value chain workforce. The discussion revealed that the majority of women are found predominantly working in the informal sector and in lower positions. This is influenced by gender-biased attitudes, stereotypes and perceptions that plastic waste activities are gender exclusive, with certain roles belonging to men while other roles belong to women.

Traditionally, women have been categorised as the cornerstones of waste management and are also predominantly in charge of domestic labour such as cleaning at home, washing and cooking and therefore the management of waste.

In the same vein when it comes to CPE, are found to adopt similar traits of collecting, cleaning and sorting – in essence putting together and making things nice compared to men. It is also noted that the cleaning services industry is 75% women. It was further outlined that, even in formal settlings such as municipalities, cleaners, e.g., road sweepers and domestics workers are mainly women.

The discussion further outlined that these gender-biased attitudes often diminish the social status of women in the communities, as they may be socially categorised and seen as doing dirty jobs. This finding collaborates with the findings in Ghana, which found that stigmatisation of women due to their involvement in plastic waste picking is high, and as a result in some communities, people dissociate themselves especially from females working in waste companies or waste pickers. (Global Plastic Action Partnership, 2021). This phenomenon is more prevalent in rural areas than is in urban settings.

3.5 The Importance of Awareness Raising and Knowledge Sharing

Discussion with academia and civil society revealed that the sphere of influence and empowerment under which women are distinguished to be dominant is environmental activism, promotion and research. Women are seen as agents of change, as they have the ability and power to convince family members, influencing attitude at the household level. Furthermore, a male lecturer at the Namibia University of Science and Technology observed that female students at his University were more likely to conduct research on waste management. The lecturer concluded that this was due to the fact that women are naturally concerned for the environment. Moreover, it has been shown that environmental activists are usually women (e.g. Nobel Prize winner in Kenya). Despite such good endeavours, all stakeholders from all the focus group discussions agreed that there is need for gender mainstreaming in order to empower women in digital innovation for CPE. Therefore, outreach should be targeted directly at women in this regard.

It is also clear from the results that women were eager to form part of groups aiming to raise awareness or information sharing on the environmental and economic values of waste management. The majority of waste management start-ups in the Namibian FGDs belongs to females which demonstrates the future control by women in the CPE, which could have significant positive future impacts in the African communities. The participation of women in the CPE was not only geared towards economic gains but had a greater focus on knowledge sharing and education of the future generation. Some stakeholders from the different African countries that participated in this study indicated programmes aiming to share transfer of knowledge to accelerate and digitise waste management among children, youth and women. This is expected to be done by clean waste campaigns, schools competitions, technology transfer, hackathons and other mindset transformational initiatives. Again, most of these initiatives were mainly

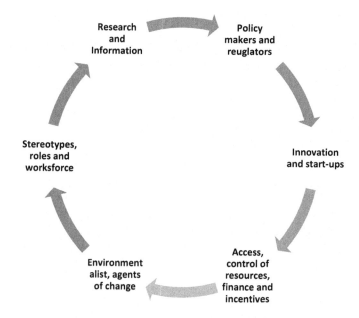

FIGURE 15.2   Persistent challenges faced by women in digital innovation in CPE in Africa

driven by women. Besides all these, the analysis highlighted different challenges faced by women in digital innovation in CPE (Figure 15.2).

4   Discussion

Gender consideration for the development of DI for CPE can be influenced by different factors. For instance, the areas to look into include supporting adequate involvement of women in decision-making processes, the need to strengthen and enhance DI in start-ups, particularly those belonging to women while allowing access of finances to women. In addition, stereotyping limits gender equality of CPE as well as equal access to knowledge and information.

The United Nations Industrial Development Organisation (UNIDO) developed a framework which includes gender-specific indicators that cover power and agency as well as economic advancement, to guide gender analysis and mainstreaming in project cycle, including waste management and DI to track outputs, outcomes and impact (Schuber, Dor & Schmidt, 2021). The framework outlined power and agency to be an essential indicator of mainstreaming gender equality. Gender-responsive action should promote progressive changes in power relationships between women and men (World Economic Forum, 2021). DI in CPE will contribute to power and agency, when there is increased participation of women in different waste management. However, DI equally takes into consideration the distinct vulnerabilities, needs and experiences of women and men (Schuber et al., 2021).

This is against the background, where there are clear patterns of women's inferior access to resources and opportunities and their under-representation in decision-making processes. The exclusion of women and marginalised groups from decision making at the highest levels of policy, operations, planning and programme design has led to a fragmented response to plastic pollution (World Economic Forum, 2021). National governments should ensure that such inequalities and differences are documented in official documentations that guide waste management and DI such as action plans, reports, frameworks, guidelines and other technical working documents. In so doing, participation in DI and CPE and benefit from capacity-building activities and awareness initiatives will impact both women and men (World Economic Forum, 2021). This will enable increased self-efficacy, bargaining power and ability to make decisions (Schuber et al., 2021). Provision of these enabling environment will result in increased financial independence and increased control of household resources (Schuber et al., 2021).

Furthermore, the framework outlines that advancements in gender equality, and the empowerment of women will also lead to improved economic advancement, consequently increasing the profit margin and improved livelihood opportunities (Schuber et al., 2021). This analysis is based on the fact that past formal economic sectors and investment favoured men compared to women due to existing inequalities (World Economic Forum, 2021). This will be achievable when DI in CPE offers women access to new markets opportunities, thus empowering start-ups by women in CPE to change their business practices and model with evolving markets, as well as provide access to learning streams education, training and skills development opportunities and transfer of environmentally sound technologies through provision of technical expertise in advancing technology use (Schuber et al., 2021; World Economic Forum, 2021). This will enable effective participation in the economic advancement by women and reduce inequalities.

Borrowing from Guyot Phung (2019), the digital innovation for CPE particularly with respect to women will be effective and successful when women are involved in sharing of ideas, and this can work most effectively when women are also involved in decision-making processes in CPE. This subsequently leads to better collaboration between women and men that can help to strengthen the know-how and take up of digital innovation which consequently leads to improved CPE in Africa for everybody. Furthermore, there is a growing recognition of the importance of digital innovation in CPE, but in order for this recognition to be meaningful, the role that women play needs to be equally recognised to the extent that their roles also need to evolve as not just mere pickers but as a group who can also play a critical role as innovators, start-ups as well as decision makers. This will need to be done in tandem with improving women's skills in the area of digital training so that they too can adequately adapt, adopt and have access to digital technologies that are constantly evolving. Lastly, Guyot Phung (2019) in his illustration covers generational needs. This is important to consider for digital innovation and CPE and what it means for the young and upcoming generation. The young, particularly young women, will need to be involved in the use of digital technologies in order to change the status quo where currently women are still lagging behind in use of and access to digital innovations. This can be effectively done through the introduction of policies, learning about digital innovations at an early age through improved curriculum in schools that encourages both boys and girls to understand and use digital innovation in equal measure. Such measures can help with the development of skills for the next generations. This is in line with the findings of this study where several programmes were organised for the school children and the youth, who are the future generation.

5 Conclusion

Even in the world of digital innovation and plastic waste, technology is not neutral and has implications on a number of aspects, including persistent gender inequalities. Typically, in as far as the African continent is concerned, this is an area still predominantly dominated by the male gender. Yet, in a world where women are mostly impacted by plastic waste, it becomes imperative to be able to recognise this challenge and subsequently find a solution in which women as much as men can use digital innovation to equally overcome the negative impact of plastic waste.

This chapter has revealed that there is some way yet to go when it comes to CPE and digital innovation because women are still under-represented in decision making which means that women do not always have presence in policy making. Furthermore, women do not have adequate financial resources due to the fact that they do not have easy access to finances that can make a significant impact to them as investors, as start-ups or have access to digital technologies as easily as men. This appears to contribute to women occupying more informal spaces as waste pickers, as street sweepers, washers and plastic sorters. Inevitably, this has health and security risks for women due to the fact that plastic waste has been proven to have hazardous chemicals as well as dangerous instruments mixed in the waste that can be harmful to the pickers. Additionally, it is unsurprising to find that male pickers will often have access to better waste picking equipment such as push carts and tricycles, something which is not often the case with women. Furthermore, men are better at negotiating better rates for themselves as pickers while the same is not the case for women. This may be due to the fact women feel they do not have the power to negotiate better rates for themselves and sometimes, even if they do, they may just be ignored. Despite the above, digital innovation promises to better women's position when they have know-how and access as it enables acquisition of new knowledge, skills and technologies necessary to improve their positions in CPE. As digital innovation is critical for improvement of household or community-level economies, it will therefore go without saying that this will be advantageous to women who are usually looking after their households as primary carers and as a result their communities. The fact that the application of digital innovation to CPE allows for smart collection and recycling of plastic waste and exchange of information in a much more effective and efficient manner will also be advantageous to women who are the main pickers. Digital innovation also presents an opportunity for CPE in the creation of job opportunities in a relatively niche area where gender inequalities still persist. Thus, this will allow for better business opportunities, acquisition of new skills and easier collaboration and networking for both men and women.

6   Recommendations

Technological development must be gender sensitive and this includes in terms of access, adaptability, affordability and use. There is also an urgent need to change the mindset on cultural norms which sees plastic waste collection as a dirty job for the poor people especially women. If this were to happen, women would be seen as equal players in CPE deserving better pay, better training and skills development in digital innovation for there to be improved CPE. Therefore, this calls for several actions which include:

- Deliberate actions and monitoring interventions by policy makers and investors
- Development and implementation of polices and legislation with gender-aggregated targets that should be promoted over gender-neutral laws currently in existence which have not been helpful in successfully bringing women at par with their male counterparts
- Coordination in efforts geared toward gender inclusiveness in different sectors – this should be at the forefront
- Need for multi-layer collaboration involving the government, businesses and civil societies so that there is no duplication of efforts so that there is knowledge exchange with all relevant stakeholders
- Need for a waste management national data base – this requires coordinated efforts and collaborations at various level
- Need for training in various technology-related areas related to plastic waste for both men and women
- Need for funding/investments and an awareness of how to access this type of funding so that both men and women benefit

The above recommendations will work well especially with policy interventions that bring all stakeholders together. Women should be clearly seen to be included and engaged in any policies and decision-making processes that are taken in as far as digital innovation and CPE in Africa is concerned.

Acknowledgement

This work was supported by the United Kingdom Research and Innovation (UKRI) Global Challenges Researches Fund (GCRF) under Grant EP/ T029846/1.

References

Ayeleru, O. O., Dlova, S., Akinribide, O. J., Ntuli, F., Kupolati, W. K., Marina, P. F., … & Olubambi, P. A. (2020). Challenges of plastic waste generation and management in sub-Saharan Africa: A review. *Waste Management, 110*, 24–42.

Baah-Boateng, W., & Twum, E. (2018). *Economic complexity and employment for women and youth: The case of Ghana.* University of Cape Town, Development Policy Research Unit, 2019. http://hdl.handle.net/10625/58423

Balcerzak P.A., Pietrzak B.M. (2017). Digital economy in visegrad countries. Multiplecriteria decision analysis at regional level in the years 2012 and 2015. *Journal of Competitiveness* 9(2).

Barefoot, K., Curtis, D., Jolliff, W., Nicholson, J. R. & Omohundro, R. (2018). Defining and measuring the digital economy; US Department of Commerce Bureau of Economic Analysis: Washington, DC, USA, 2018. Available online: www.bea.gov/digitaleconomy/_pdf/defining-and-measuring-the-digital-economy.pdf (accessed on 20 July 2021).

Berg, H. & Wilts, H. (2019, March). Digital platforms as market places for the circular economy—requirements and challenges. In *Nachhaltigkeits Sustainability Management Forum* 27 (1), 1–9. Springer Berlin Heidelberg. DOI: 10.1007/s00550-018-0468-9

Bukht, R. & Heeks, R. (2018). Defining, conceptualizing and measuring the digital economy. Development Informatics working paper. *International Organisations Research Journal* 13, 143–172.

Desmond, P. & Asamba, M. (2019). Accelerating the transition to a circular economy in Africa: Case studies from Kenya and South Africa. In *The Circular Economy and the Global South* (pp. 152–172). Routledge.

European Environment Agency (2021). Digital technologies will deliver more efficient waste management in Europe. Accessed from www.eea.europa.eu/themes/waste/waste-management/digital-technologies-will-deliver-more

Global Plastic Action Partnership (2021). Gender analysis of the plastics and plastic waste sectors in Ghana Baseline Analysis Report, May 202: Accessed from www.wacaprogram.org/sites/waca/files/knowdoc/NPAP-Ghana-Gender-Baseline-May-2021.pdf.

Godfrey, L., Nahman, A., Yonli, A. H., Gebremedhin, F. G., Katima, J. H., Gebremedhin, K. G., … & Richter, U. H. (2018). Africa waste management outlook. https://wedocs.unep.org/bitstream/handle/20.500.11822/25515/Africa_WMO_Summary.pdf?sequence=1&isAllowed=y

GSMA. (2021). Digital Dividends in Plastic Recycling: Accessed from www.gsma.com/mobilefordevelopment/wp-content/uploads/2021/04/ClimateTech_Plastic_R_WebSingles2.pdf

Guyot Phung, C. (2019). Implications of the circular economy and digital transition on skills and green jobs in the plastics industry. *Field Actions Science Reports. The Journal of Field Actions* (Special Issue 19), 100–107.

Joshi, C., Seay, J. & Banadda, N. (2018). A perspective on a locally managed decentralized circular economy for waste plastic in developing countries. *Environmental Progress & Sustainable Energy, 38*(1), 3–11.

Kaur, G., Uisan, K., Ong, K. L. & Ki Lin, C. S. (2017). Recent trends in green and sustainable chemistry & waste valorisation: Rethinking plastics in a circular economy, *Current Opinion in Green and Sustainable Chemistry*, 9, 30–39. https://doi.org/10.1016/j.cogsc.2017.11.003

Khan, S., Khan, S. & Aftab, M. (2015). Digitization and its impact on economy. *International Journal of Digital Library Services, 5*(2), 138–149.

Longwe, S.H. (2002). Addressing Rural Gender Issues: A Framework for Leadership and Mobilisation. Paper presented at the III World Congress for Rural Women, Madrid.

March, C, Smyth, I., & Mukhopadhyay, M. (1999). A guide to gender – analysis framework, 1999, available at www.ndi.org/files/Guide%20to%20Gender%20Analysis%20Frameworks.pdf

Mourdoukoutas, E. (2017). Africa's digital rise hooked on innovation. United Nations Africa Renewal. *Africa Renewal, 31*(1), 34–35. Retrieved from www.un.org/africarenewal/magazine/may-july-2017/africa%E2%80%99s-digital-rise-hooked-innovation

Nandi, S., Sarkis, J., Hervani, A. A. & Helms, M. M. (2021). Redesigning supply chains using blockchain-enabled circular economy and COVID-19 experiences. *Sustainable Production and Consumption, 27*, 10–22.

Ocean Conservancy. (2019). *The Role of Gender in Waste Management: Gender Perspectives on Waste in India, Indonesia, the Philippines and Vietnam.* Singapore: Ocean Conservancy.

OECD. (2020). Gender-specific consumption patterns, behavioural insights, and circular economy. Mainstreaming gender and empowering women for environmental sustainability. Retrieved from www.oecd.org/env/GFE-Gender-Issues-Note-Session-5.pdf

Oliveira, T. A., Oliver, M. & Ramalhinho, H. (2020). Challenges for connecting citizens and smart cities: ICT, e-governance and blockchain. *Sustainability, 12*(7), 2926.

O. Nyumba, T., Wilson, K., Derrick, C. J., & Mukherjee, N. (2018). The use of focus group discussion methodology: Insights from two decades of application in conservation. *Methods in Ecology and Evolution, 9*(1), 20–32.

Park, H., & Choi, S. O. (2019). Digital innovation adoption and its economic impact focused on path analysis at national level. *Journal of Open Innovation: Technology, Market, and Complexity, 5*(3), 56.

Portulans Institute (2021). Africa's Digital Economy: Opportunities and Obstacles during COVID-19.

Rymarczyk, J. (2020). Technologies, opportunities and challenges of the industrial revolution 4.0: theoretical considerations. *Entrepreneurial Business and Economics Review, 8*(1), 185–198.

Schmitz H., 2015. Africa's biggest recycling hub? Institute of Development Studies Blog. Available at: www.ids.ac.uk/opinion/africa-s-biggest-recycling-hub

Schröder, P., Lemille, A. & Desmond, P. (2020). Making the circular economy work for human development. *Resources, Conservation and Recycling, 156*, 104686.

Schuber, C., Dor, D. & Schmidt, N. (2021). Guide to gender analysis and gender mainstreaming the project cycle. Retrieved from www.unido.org/sites/default/files/files/202106/Gender_mainstreaming_Guide_1_Main%20guide.pdf

Shanahan, M. & Rowntree, O. (2019). Realising the full benefit of mobile for women in Africa. Accessed from www.gsma.com/mobilefordevelopment/blog/realising-the-full-benefit-of-mobile-for-women-in-africa/, on 04.10.2021.

Stahel, W. (2016). The circular economy. *Nature*, 531, 435–438. https://doi.org/10.1038/531435a

United Nations, (2018). Technology and Innovation Report: Harnessing Frontier Technologies for Sustainable Development. United Nations Conference on Trade and Development, Switzerland.

Wilson, M., Kitson, N., Beavor, A., Ali, M., Palfreman, J. & Makarem, N. (2021). Digital Dividends in Plastic Recycling, GSMA, UK. www.gsma.com/mobilefordevelopment/wpcontent/uploads/2021/04/ClimateTech_Plastic_R_WebSingles2.pdf

World Bank (2021). World Bank's digital economic for Africa Country Diagnostics Status Report, accessed from www.worldbank.org/en/programs/all-africa-digital-transformation/country-diagnostics, on 04.10.2021.

World Economic Forum. (2021). Guide to ensure gender-responsive action in eliminating plastic pollution. Retrieved from https://globalplasticaction.org/wp-content/uploads/GPAP-Global-Gender-Guidance-May-2021.pdf

Yoo, Y., Henfridsson, O. & Lyytinen, K. (2010). Research commentary—the new organizing logic of digital innovation: an agenda for information systems research. *Information Systems Research*, *21*(4), 724–735.

Yousaf, Z., Radulescu, M., Sinisi, C. I., Serbanescu, L. & Păunescu, L. M. (2021). Towards sustainable digital innovation of SMEs from the developing countries in the context of the digital economy and frugal environment. *Sustainability*, *13*(10), 5715.

16

CONCLUSION

The future of digitisation for the circular plastic economy in Africa

Oluwaseun Kolade and Muyiwa Oyinlola

1 When the digital meets the circular

Globally, the conversation about the circular economy has gathered momentum in recent years. Increasing frequencies of destructive climate disasters such as the East African drought, Australian wildfires, cyclones and South Asian floods have focused minds about the urgent imperative of the sustainability agenda. At the heart of the global conversation is the need to fundamentally rethink the way humans use limited natural resources for production and the need to embrace new habits of consumption that cut waste and optimise value. Circularity became the most effective catchword for this campaign. Within the past decades, thousands of scholarly articles and policy papers on the circular economy have been produced, and the circularity conversation has gained traction in academic and practitioner conferences, as well as public fora and social media.

Within the wider conversation about the circular economy, plastic pollution has taken a central stage. This is because plastic products constitute a peculiarly problematic challenge, among other manufactured products. Plastics are non-biodegradable, generally cheaper to produce, and are typically made of composites including coolants and adhesives that make them difficult to recycle. Once disposed indiscriminately, they find their way into the world's oceans where they are a great menace to aquatic life and to the wider ecosystem. The circular plastic economy is therefore an effort to challenge and reshape the linear trajectory of plastic production and consumption to a model that transforms the entire plastic value chain from design phase to production-and-use phase all the way to the end-of-life phase (Johansen et al., 2022). This transformational process is underpinned by the philosophy of resource efficiency and value optimisation realised through innovation and creativity.

DOI: 10.4324/9781003278443-19

This is where the circular meets the digital. As discussions about the circular economy gathers pace, talks about digital transformation has reached feverish pitch – not least in the wake of the Covid-19 pandemic. With heightened interests come greater scrutiny of the hitherto prevailing orthodoxy of digital innovations. Thus, in recent years, more stakeholders are questioning whether digital technologies should be about maximising production or whether digital innovations should, in fact, prioritise the pressing human challenges around pollution control and sustainable use of limited natural resources. Industry 4.0 is therefore being transcended by Industry 5.0, the latter built on three key pillars of humancentricity, resilience and sustainability (Xu et al., 2021). Humancentricity emphasises the need to prioritise human needs in the use of technology. In other words, increasing production should not be an end in itself. Rather, the impact of digital innovations should be measured in terms of the functional outcomes of the innovations in relation to human needs (Kolade and Owoseni, 2022). The second pillar of Industry 5.0, resilience, refers to applications of digital innovations to overcome disruptions and turbulence in relation to economic and social systems (Sindhwani et al., 2022). The sustainability strand focuses attention on resource efficiency. In other words, it highlights the need to apply Industry 4.0 innovations in new and creative ways to optimise limited natural resources both in relation to production systems and consumption habits (Breque et al., 2021). Thus, Industry 5.0 is not conceived as a linear progression of Industry 4.0 with, for example, the emergence of new technologies previously unknown in the Industry 4.0 phase. Rather, it is a paradigm shift that invites stakeholders to find creative ways of applying existing and emerging technologies to tackle pressing global challenges and human needs.

The circular plastic economy is a prime example of an area where Industry 5.0 principles can make a big impact. It fits within the wider conversation about the sharing economy and collaborative consumption. Digital technologies are being used to link stakeholders, mobilise new actors, and empower existing actors to actively engage in a win–win process to drive a transition to a circular plastic economy. Nowhere is this more important than the African continent, where the industrial process is still in the early stages, and the economy is not locked into old technologies and sunk investment associated with linear economy paradigm. Instead, African countries are presented with an auspicious opportunity to leapfrog and lead the world in greener technologies that do not, in the same breadth, compromise growth ambitions.

In many ways, the huge opportunities for the African continent to play a leading role in the circular economy agenda is a key premise of this book. Across its 16 chapters, scholars, policymakers and practitioners have woven together a compelling narrative that illuminates the challenges as well as opportunities for a circular plastic economy on the continent. Digital innovations are at the heart of this, the linchpin of an ambitious agenda to create and invigorate new ecosystems powered by forward-looking policies, platform innovations and new production systems. The chapters also highlight the nuances and peculiarities of country-specific contexts. The followings are the key takeaways from the book.

2 Key takeaways

Digital innovations are essential for inter-sectoral linkages and stakeholder collaborations

The chapters in this book highlight a number of exciting innovations and high-impact initiatives driven mostly by startups and third sector organisations across the continent. However, it is clear that many actors continue to work in silos, the consequences of which is that the full potentials of these initiatives are not fully realised (Kolade et al., 2022; Oyinlola et al., 2022). At the same time, in the public sector, national governments are showing increasing commitment to tackle the challenge of plastic pollution through ambitious regulations and policies. However, many of these policy interventions have followed a top-down approach, with minimal engagement with the private sectors and frontline non-governmental organisations NGOs.

This book highlights the potentials of digital innovations to facilitate a new area of inter-sectoral engagement and collaborative synergy. For example, as discussed in several chapters, platform innovations are being deployed by digital innovators to link up waste collectors with recyclers in a process that creates new income opportunities for poorer households and otherwise invisible informal operators, while reducing logistics and transaction costs for recyclers. However, when big corporations connect with these multisided digital platforms, the impact of the ecosystem is multiplied via investment in recycling centres and operators and via a digitally enabled deposit refund system. The impact is even greater with the involvement and support of the public sector, through strategic procurement policies by which governments drive demand for recycled products, tax incentives for innovators, and direct investment in digital platforms.

Among others, this book highlights the immense and under-utilised potentials of blockchain technologies. Blockchain solutions are being trialled in a number of African countries, including South Africa. In the example of BanQu in South Africa, blockchains technology has been used to mobilise waste reclaimers and recyclers in a campaign sponsored by Coca Cola, a major corporate manufacturer of plastic packaging. While the effort has achieved considerable success, there is little evidence of government involvement in the project. This exemplifies a critical missing gap across the continent, where a lot of promising initiatives have been launched independent of national governments, to considerable local successes but ultimately limited outcomes in scaling. In order to scale and sustain innovative campaigns on national and continental levels, governments should be connected and actively engaged on these platforms.

Another key area discussed in this book is data sharing. This is one of the biggest impediments to circular economy transition on the continent. Digital innovators often work in the dark in the development of digital products for the circular economy. Similarly, a lot of policy making on the continent are hampered by limited and inadequate data, for example, about consumption behaviour and disposal practices of households and organisations. A plastic data exchange, proposed in this book, can be pathbreaking for stakeholder contributions and impact for the circular plastic economy. It will enable digital innovators to create and continually improve higher-value products for other stakeholders engaged in the ecosystem. It will also help policy makers to launch targeted and more efficient interventions aimed at specific sub-sectors and players in the circular plastic ecosystem.

Curiously, there are intimations of technological scepticism from some national governments which are probably reluctant to play catch up or follow the lead of processes and campaigns initiated by startups. Attitudes are gradually changing, however, with some governments, including Nigeria, Kenya and Rwanda recently setting up ministries and agencies for "digital economy". In many cases, government agencies are now convening, as well as participating, in initiatives relating to a digitally enabled circular economy.

The potentials and networking effects of digital innovations are not limited to platform innovations and blockchain solutions. There are also opportunities for digital applications in plastic production and re-manufacturing processes. One good example discussed in this book is 3D printing, a disruptive, additive manufacturing process that upends the traditional logistics and inventory requirements of traditional manufacturing. Three-dimensional printing offers the opportunities for plastic wastes to be repurposed as pellets in re-manufacturing of plastic and composite products. It also enables the economy of one in place of the economy of scale, thereby bringing smaller operators and microenterprises into play. Moreover, its digital foundation offers opportunities for various stakeholders to contribute, including waste collectors, programmers and operators of 3D printers, as well as governments. The latter point brings us to the next key point raised in this book: the critical role of government in a digitally enabled circular plastic ecosystem.

Digital innovations must be matched with policy innovations and political will

While digital innovations offer a wide range of exciting prospects for the circular plastic economy, the book also highlights the urgent imperative of policy innovations and political commitment to the circularity agenda. The emerging evidence from countries like Rwanda is that when policies and regulations are in sync with digital innovations, much greater success is achieved for the circular plastic economy. This point aligns with the broader theme of inter-sectoral linkages and stakeholder synergy that cut across the chapters in the book: that ideas work better when linked with other ideas, and stakeholders can achieve much greater impact collectively than the sum of their individual contributions.

Thus, the first requirement on this front is that African countries need new ideas to accelerate the transition to a circular plastic economy. In order to generate new ideas, government and public sector countries must look outwards for examples of best practices across the world, where circular economy policy ideas have been successful. As the old saying goes, African countries do not have to re-invent the wheel of circular plastic policies. Yet, African countries must do more than imitate and adopt policy ideas from oversees, not least because the institutional, political and cultural contexts are different even among individual African countries, not to mention the differences in relation to non-African contexts. Therefore, circular policy innovations in Africa should be achieved, not by wholesale importation of new ideas, but by aggregation, integration and customisation of new ideas to meet specific needs and outcomes in individual countries.

This is where policymakers must look as much within as they look outside their countries. Since policies are made for people, rather than people for policies, national governments should leverage on the existing ecosystem and networks of non-state actors in the circular plastic economy. These include digital innovators and frontline non-governmental organisations whose knowledge and experience can give national governments access to critical data and information about the local contexts in which policies are being enacted and implemented. These frontline organisations and enterprises can also act as the critical bridge between governments and households and communities in the grassroots, in terms of policy formulation, communication and implementation. More often than not, policies fail because they are disconnected from the realities of everyday life and because of low levels of citizens engagement. In order to mitigate this, policymakers should work with frontline civic organisations and digital enterprises to engage citizens from the early stages of policy ideation and formulation, all the way to implementation. Community co-ownership of policies and regulations is essential to the success of circular plastic economy in African countries.

The point on community ownership of public policies should be complemented with an equally important point about political will. While the challenges of sunk investment and vested interests are relatively light in relation to the linear economy on African continent, the influence of multinational corporations and major manufacturers is disproportionately big in many African countries. In some cases, big corporations have forged alliance with certain sections of the political elite to achieve state capture, rooted in an unwritten commitment to existing linear paradigm of production and pollution patterns. In order to upend this trajectory, Africa needs strong political leadership and commitment to circular plastic economy ideas. This is required, for example, to make the use of virgin products less desirable for the manufacture of plastic products. As this book highlights, Rwanda exemplifies this type of strong leadership and commitment to the circularity agenda. There, the national leadership has shown its ability to take a stand, sometimes against big corporate interests, to drive circular, environment-friendly policies. Conversely, Rwanda also seems to expose a weakness in having a strong political leadership that is not adequately complemented with civic and community engagement. Cultures and attitudes are changing more slowly than policies and regulations. You need a good blend of strong leadership and public ownership to achieve long-term impact, institutional transformation and cultural change beyond the tenures of specific governments in power. You need this combination to achieve a market-driven circular plastic economy which can run and grow on its own with only light-touch interventions from national governments. We now turn attention to this third and final point.

Digital innovation is the engine of a market-driven circular economy in Africa

If the circular plastic economy is to gain enough traction to upend the trajectory of linear production and consumption, it needs to be market-driven. Activist government interventions and regulations are required, especially at the initial phases, to create demands and invigorate the market. As the market grows and more stakeholders become involved, government involvement need to focus less on regulations and penalties and more on incentives and market-oriented mechanisms such as strategic public procurement. One of the main challenges with incentives and market-driven interventions is that they often miss their targets in an environment where institutions are relatively weak and partisan politics dominate the implementation of public policies. In such politically charged environments, financial and in-kind incentives from governments are sometimes used to reward political party loyalists or channelled via party activists to consolidate political power. In these environments, typical of many African countries, digital innovations can be harnessed for transformative market impacts in three key areas.

Firstly, digital innovations can be used to ensure transparency and greater effectiveness of market-oriented public policies. In a digitally mediated, publicly accessible market-oriented intervention, stakeholders can be easily enabled to track and monitor front-end implementations of public policies. The public will be able to see, for example, which incentives are going to which organisations and actors in the circular plastic ecosystem. This can significantly cut corruption and waste. For service-oriented and outcome-driven governments, the digital transformation of public policies can be harnessed as a positive sum game in which governments in power realise political capital from a transparent, traceable implementation of market-based interventions. Publicly visible impacts of policies can be used to win hearts and minds.

Secondly, digital innovations provide circular economy actors with platforms and opportunities to better organise themselves, not only to access benefits from existing market interventions but also to influence future policies to scale and grow the market for the circular plastic economy. In many African countries, this organisation is at very early stages, with big gaps and opportunities for impact and growth. Even more, there are huge market opportunities at cross-country and continent-wide levels which can be realised with digitally enabled organisation of market actors. This book highlights reflections from stakeholders in focus groups and in-depth interviews about these opportunity gaps and the potentials they offer for the future of circular plastic economy on the continent.

Finally, digital innovations are essential for increased and easier access of the general public to circular plastic products. There is a huge market for circular products on the African market. Some of this is partly related to the paradox of income inequality on the continent. Currently, a considerable proportion of multiple-use plastic products disposed by mid- and high-income households are usable and in high demand in low-income communities. Digital innovations can directly link those disposing items with those who need them as is or connect them with recyclers. Similarly, digital platforms can be used to connect buyers with sellers of remanufactured plastic products. The deployment of digital innovations helps, among others, to simplify the logistics of access and the transaction costs normally associated with market processes.

3   Final thoughts

This book has highlighted a range of ecosystem, institutional and market opportunities associated with a digitally enabled circular plastic economy in Africa. It did not shy away from the challenges, either. The circular plastic economy is promising and realisable only if stakeholders work together, if national governments summon the will and wits and market processes are enabled to drive the circular transition. Digital platforms are particularly effective in linking stakeholders together for synergistic collaboration, both in virtual and face-to-face settings. In the public sector, digital products work much better with government buy ins and can enable access to otherwise difficult to access data needed for effective public policy. Finally, digital innovations are essential for the circular plastic economy agenda to work as a viable, profitable business. Digital innovations, including multisided digital platforms, can drive circular business models. They can also be used to scale successful market-oriented interventions through the inclusion and empowerment of new actors in the circular plastic ecosystem.

References

Breque, M., de Nul, L., Petrides, A., 2021. Industry 5.0 – Towards a sustainable, human-centric and resilient European industry, European Commission. https://doi.org/10.2777/308407

Johansen, M.R., Christensen, T.B., Ramos, T.M., Syberg, K., 2022. A review of the plastic value chain from a circular economy perspective. J Environ Manage 302, 113975. https://doi.org/10.1016/j.jenvman.2021.113975

Kolade, O., Odumuyiwa, V., Abolfathi, S., Schröder, P., Wakunuma, K., Akanmu, I., Whitehead, T., Tijani, B., Oyinlola, M., 2022. Technology acceptance and readiness of stakeholders for transitioning to a circular plastic economy in Africa. Technol Forecast Soc Change 183. https://doi.org/10.1016/j.techfore.2022.121954

Kolade, O., Owoseni, A., 2022. Employment 5.0: The work of the future and the future of work. Technol Soc 71, 102086. https://doi.org/10.1016/j.techsoc.2022.102086

Oyinlola, M., Schröder, P., Whitehead, T., Kolade, O., Wakunuma, K., Sharifi, S., Rawn, B., Odumuyiwa, V., Lendelvo, S., Brighty, G., Tijani, B., Jaiyeola, T., Lindunda, L., Mtonga, R., Abolfathi, S., 2022. Digital innovations for transitioning to circular plastic value chains in Africa. Africa J Manage 8, 83–108. https://doi.org/10.1080/23322373.2021.1999750

Sindhwani, R., Afridi, S., Kumar, A., Banaitis, A., Luthra, S., Singh, P.L., 2022. Can industry 5.0 revolutionize the wave of resilience and social value creation? A multi-criteria framework to analyze enablers. Technol Soc 68, 101887. https://doi.org/10.1016/j.techsoc.2022.101887

Xu, X., Lu, Y., Vogel-Heuser, B., Wang, L., 2021. Industry 4.0 and Industry 5.0—Inception, conception and perception. J Manuf Syst 61, 530–535. https://doi.org/10.1016/j.jmsy.2021.10.006

INDEX

Printed in the United States
by Baker & Taylor Publisher Services